COMPUTATIONAL ECONOMICS AND ECONOMETRICS

Advanced Studies in Theoretical and Applied Econometrics

Volume 22

The titles published in this series are listed at the end of this volume.

Computational Economics and Econometrics

Edited by

Hans M. Amman
David A. Belsley
Louis F. Pau

SPRINGER SCIENCE+BUSINESS MEDIA, B.V.

Library of Congress Cataloging-in-Publication Data

Computational economics and econometrics / edited by Hans M. Amman,
David A. Belsley, Louis F. Pau.
p. cm. -- (Advanced studies in theoretical and applied
econometrics ; v. 22)
Includes index.
ISBN 978-94-010-5394-5 ISBN 978-94-011-3162-9 (eBook)
DOI 10.1007/978-94-011-3162-9
1. Econometrics--Congresses. 2. Computer simulation--Congresses.
I. Amman, Hans M. II. Belsley, David A. III. Pau, L.-F. (Louis
-François), 1948- IV. Series.
HB139.C645 1991
330'.01'5195--dc20 91-16876

ISBN 978-94-010-5394-5

Printed on acid-free paper

TABLE OF CONTENTS

Preface vii

PART ONE: ECONOMETRICS

Likelihood evaluation for dynamic latent variables models
David F. Hendry and Jean-François Richard 3

Global optimization of statistical functions: Preliminary results
William L. Goffe, Garry D. Ferrier and John Rogers 19

On efficient exact maximum likelihood estimation of high-order multivariate ARMA models
Stefan Mittnik 33

Efficient computation of stochastic coefficients models
I-Lok Chang, Charles Hallahan and P.A.V.B. Swamy 43

The degree of effective identification and a diagnostic measure for assessing it
David A. Belsley 55

PART TWO: MODEL STIMULATION AND OPTIMIZATION

A splitting equilibration algorithm for the computation of large-scale constrained matrix problems: Theoretical analysis and applications
Anna Nagurney and Alexander Eydeland 65

Nonstationary model solution techniques and the USA algorithm
P.G. Fisher and A.J. Hughes Hallett 107

Implementing no-derivative optimizing procedures for optimization of econometric models
György Barabás 121

Information in a Stackelberg game between two players holding different theoretical views: Solution concepts and an illustration
Henriëtte M. Prast 137

Exchange rate uncertainty in imperfect markets: A simulation approach
Hans M. Amman and Lidwin M.T. van Velden 157

Subject index 165

Authors' index 169

PREFACE

The field of *Computational Economics* is a fast growing area. Due to the limitations in analytical modeling, more and more researchers apply numerical methods as a means of problem solving. In turn these quantitative results can be used to make qualitative statements. This volume of the *Advanced Series in Theoretical and Applied and Econometrics* comprises a selected number of papers in the field of computational economics presented at the Annual Meeting of the Society Economic Dynamics and Control held in Minneapolis, June 1990.

The volume covers ten papers dealing with computational issues in Econometrics, Economics and Optimization. The first five papers in these proceedings are dedicated to numerical issues in econometric estimation. The following three papers are concerned with computational issues in model solving and optimization. The last two papers highlight some numerical techniques for solving micro models.

We are sure that *Computational Economics* will become an important new trend in Economics in the coming decade. Hopefully this volume can be one of the first contributions highlighting this new trend.

The Editors

H.M. Amman et al. (eds), Computational Economics and Econometrics, vii.

PART ONE

ECONOMETRICS

LIKELIHOOD EVALUATION FOR DYNAMIC LATENT VARIABLES MODELS[1]

DAVID F. HENDRY
Nuffield College, Oxford, U.K.
and
JEAN-FRANÇOIS RICHARD
ISDS, Pittsburgh University, Pittsburgh, PA, U.S.A.

Abstract. We propose a general Monte Carlo simulation technique for evaluating the likelihood function of dynamic latent variables models, based on artificial factorizations of the sequential joint density of the observables and latent variables. The feasibility of the proposed technique is demonstrated by means of a pilot application to a one-parameter disequilibrium model. Extensions to models with weakly exogenous variables and the use of acceleration methods are discussed.

1. Introduction

Since the 1970s there has been a major resurgence of interest in the topic of Dynamic Latent Variables (DLV) models. For the purpose of the present discussion, latent variables are broadly defined as variables which enter the formulation of an econometric model and yet are not observable. See in particular the discussion in Aigner *et al.* (1983). Latent variables are widely recognized to be major components of the behavior of economic agents. They are inherently dynamic for a broad class of models such as intertemporal optimization, search or duration processes, error correction mechanisms, models of habit formation or persistence, and state dependence. Important examples are discrete choice models (Heckman, 1981) and disequilibrium models (Quandt, 1988; references in Quandt, 1989).

Unfortunately, with the exception of linear Gaussian models, the likelihood functions of DLV models are often analytically intractable, largely because the elimination of the latent variables requires high-dimensional numerical integration. Many important techniques have been developed over recent years that address the specific issue of numerical tractability. For expository purposes we can usefully regroup these contributions into three broad categories (not mutually exclusive):

1. *Simplifying the dynamic structure.* A characteristic example of this approach is the use of autocorrelated error terms in the (static) supply and demand equations of disequilibrium models. In Laffont and Montfort (1979), the resulting likelihood function leads to direct numerical evaluation of the maximum likelihood estimators.

[1] Financial support for this work has been provided by the Ford Foundation, the National Science Foundation (SES–9012202) the Pew Charitable Trust, and by the UK Economic and Social Research Council under Grants B0022012 and R231184. We are pleased to acknowledge useful comments from J. Danielsson and J. Geweke. The usual disclaimer applies.

H.M. Amman et al. (eds), Computational Economics and Econometrics, 3–17.

2. *Observable surrogates for latent variables.* Typical examples are the use of price-adjustment equations in disequilibrium models (see Quandt (1982, 1988) or Maddala (1983)) or the use of "fixed-effects" formulations as in Heckman and McCurdy (1980).

3. *Simulation.* Even though the likelihood functions associated with DLV models may be intractable, the models themselves are often amenable to *joint* simulation of the latent and observable processes. Hence the recently developed Method of Simulated Moments (MSM), as discussed e.g. by McFadden (1989) or Pakes and Pollard (1989), is applicable to DLV models and is discussed further below.

Each line of attack has its merits. The first often results in more manageable formulations. The second offers the major advantage that the inclusion of sensible surrogates generates additional information. The simulation techniques may be computationally more demanding, but they are extremely general and therefore have considerable potential for preserving the full dynamic structure of the problem under consideration.

In the present paper we aim at evaluating the likelihood function *itself* by means of simulation techniques. The motivation for doing so is obvious: a broad range of statistical techniques (estimation, hypothesis testing, Bayesian methods) is likelihood based. The difficulty to be addressed lies in the fact that the joint distribution of the latent variables *conditional* on the observables is generally not available for DLV models. On then has to select an approximating distribution, known as an *importance function*, in the form of a random number generator. The main object of our paper is to propose a general procedure for the automatic selection of importance functions and demonstrate its feasibility with a pilot application. This paper describes work in progress, and only briefly considers the design of "variance reduction" or "acceleration" techniques, and extensions to models with unmodelled variables. The reader can also usefully refer to the current literature on Monte Carlo integration (see e.g. Geweke (1989) for many technical details and for additional references).

This paper is organized as follows: Section 2 considers the closely related literature on MSM; the automatic selection of importance functions is discussed in Section 3, and an application is proposed in Section 4. Acceleration techniques are noted in Section 5. The treatment of exogenous variables is examined in Section 6, and Section 7 concludes.

2. Method of Simulated Moments

The objective of this section is to describe, and relate our own work to, current developments, in particular to the Method of Simulated Moments (MSM). For ease of presentation, we temporarily abstract from specifically discussing the *sequential* (dynamic) structure of the class of models under consideration and adopt instead a global point of view. Let y denote a vector of observables and s a vector of latent variables. Let $f(y, s|\theta)$ denote the joint distribution of y and s, where θ is a vector of unknown parameters. The likelihood function of the observables, $L(\theta; y)$, is then given by:

$$L(\theta; y) \propto \int f(y, s|\theta)\, ds\,. \tag{1}$$

Analytical integration in (1) is generally intractable for DLV models. The same applies to conventional numerical integration (quadratures, Gaussian rules, etc.) due to the inherent dimensionality of the problem — typically a multiple of sample size. In contrast, simulation techniques have been applied with considerable success in high-dimensional problems. Their application to (1) for any given value of θ requires that a random number generator be available for s, *y being held fixed*. This would obviously be the case if the distribution of (y, s) given θ could be factored into the form

$$f(y, s|\theta) = f_a(s|y, \theta_a) \cdot f_b(y|\theta_b)\,. \tag{2}$$

Unfortunately, DLV models are generally specified in the form of a *sequential* dynamic process, whereby variables are generated at time t conditionally on the past history of the process. With few exceptions[2], the transformation from such a sequential factorization of the model into a global distribution for s given y is largely intractable.

This may be the major reason that direct evaluation of (1) by Monte Carlo techniques is largely held to be intractable (see McFadden (1989) and Pakes and Pollard (1989)). Before addressing this issue, we introduce MSM in this context.

Since random number generators are typically available for the sequential distribution of y and s, they can also be used to draw random samples from their joint distribution. Hence integrals of the form

$$H(\theta) = \int \int h(y, s, \theta) \cdot f(y, s|\theta)\, ds\, dy \tag{3}$$

(where $h(\cdot)$ is some appropriate function) can be estimated by simulation for a broad range of complicated dynamic models. It is well known that integrals of the form (3) are instrumental in the derivation of generalized-method-of-moments estimators (see Hansen, 1982). Furthermore, along the lines of the discussion in Hendry (1976) in the context of simultaneous-equations models, (3) can be construed as an Estimator Generating Equation (EGE) leading to the derivation of a broad range of estimators. Specifically, let θ_0 denote the true value of θ. The statistical formulation of a model naturally suggests choices of $h(\cdot)$ for which $H(\theta_0)$ equals zero. Typical

[2] Obvious exceptions are:
(i) linear Gaussian models, for which the joint distribution of y and s is also Gaussian;
(ii) models which are such that the latent variables do not feedback onto the observables (a concept known in econometrics as that of "non-causality" in the sense of Granger, 1969).

examples are orthogonality conditions and score vectors. Alternatively, $h(\cdot)$ can be any function such that a point estimate, $\hat{H}$, is available for $H(\theta)$.

Overall, MSM consists of using simulation techniques to evaluate $H(\theta)$ for different values of θ. A value of θ is then selected that minimizes a suitable distance between the estimate $\hat{H}$ and the predicted value $H(\theta_0)$. This line of attack has proved successful (see in particular Lerman and Manski (1981), Pakes (1986), Hajivassiliou (1989), Mc Fadden (1989) and Hajivassiliou and McFadden (1989)). Also, major theoretical progress has been achieved in the investigation of the properties of these simulated estimators (see Andrews (1989), Duffie and Singleton (1989), and Pakes and Pollard (1989)).

The major attraction of MSM lies in its conceptual simplicity and ease of adaptation to a wide range of sophisticated models. Its application to integrals of the form in (1) requires, however, that y be kept fixed in the simulation which, as already discussed, raises an additional layer of complexity that, to the best of our knowledge, has not been addressed in the context of DLV models.

A recent contribution by Hajivassiliou and McFadden (1989) in the context of Limited Dependent Variables (LDV) models constitutes an original attempt at conditioning on the observables, which is precisely our own objective. The authors consider a situation where (in our notation) the observable y is a discrete, many-to-one, transform τ of the latent variable s. The likelihood of y is then given by

$$L(\theta; y) = \int_{A(y)} f(s|\theta)\, ds\,, \tag{4}$$

where $A(y) = \{s|y = \tau(s)\}$, and could be estimated by the sampling frequency of the event $A(y)$.[3] Conceptually at least, we could extend this technique to approximate the integrals in (1) at any given y_0 by limiting-sequences of integrals of the form

$$L(\theta; y_0) = \lim_{\varepsilon \to 0} \frac{\text{Prob}\,(N_\varepsilon(y_0))}{\text{Vol}\,(N_\varepsilon(y_0))}\,, \tag{5}$$

$$\text{Prob}\,(N_\varepsilon(y_0)) = \int_{N_\varepsilon(y_0)} f(y, s|\theta)\, dy\, ds\,, \tag{6}$$

where $N_\varepsilon(y_0)$ denotes an ε-neighborhood of y_0. Practical problems would probably arise regarding the choice of an appropriate sequence of ε's in (5) and the existence

[3] Naturally a "brute force" frequency count might prove extremely inefficient since the probability of the event $A(y)$ is typically (very) small. Hence the authors do not evaluate the likelihood function itself but construct instead a random number generator based on Gibbs resampling technique, as discussed e.g. in Geman and Geman (1984), that enables them to draw directly from the distribution of s, *truncated* to the event $A(y)$. These random numbers are then used to construct a Monte Carlo estimate of the score function.

of a "numerical" limit. This is why we instead propose to construct importance functions meant to approximate the conditional distribution of s given y, and to do so in a way which is applicable to a large class of DLV models.

3. Simulated Likelihood Function

We now specifically account for the dynamic structure of the models under consideration. Let y_t denote a vector of random variables observable at time t and s_t a vector of unobservable or latent variables. A sample of size T is available for y_t in the form of a matrix Y. The corresponding matrix S has not been observed. The notation Y_{t-1} and S_{t-1} represents the matrices $(y_1, \dots, y_{t-1})$ and $(s_1, \dots, s_{t-1})$, respectively. Let $f(Y, S)$ denote the joint data density for Y and S, where parameters and exogenous variables are omitted for ease of notation. In what follows, we shall introduce diverse factorizations of this joint data density, all factors of which are generically denoted by $f(\cdot|\cdot)$. Initial conditions are either known or included in S, depending upon the context. As in (1), the likelihood function of interest is proportional to the marginal data density

$$f(Y) = \int f(Y, S)\, dS\,, \tag{7}$$

where the dimensions of integration can be quite high (the simple disequilibrium model we mention below necessitates a 50 dimensional integration). The crucial step in estimating (7) by Monte Carlo techniques consists in factoring the joint density $f(Y, S)$ into the product of an importance function $g_Y(S)$ — from which samples can be drawn at random — and a remainder function $h(Y, S)$, which is implicitly defined by the identity

$$f(Y, S) = h(Y, S) \cdot g_Y(S)\,. \tag{8}$$

The notation $g_Y(\cdot)$ is meant to emphasize two critical aspects of the proposed methodology.

1. $g_Y(\cdot)$ need not, and in general will not, represent a "valid" reduction of $f(Y, S)$. In particular, $g_Y(\cdot)$ need not coincide with the actual conditional sampling distribution $f(S|Y)$. There are very few applications for which the latter is available in the form of an operational random number generator. Considerable flexibility is gained by allowing $g_Y(\cdot)$ to be *any suitable auxiliary distribution* constructed for the sole purpose of evaluating the integral in (7).

2. Throughout the integration, the Y's are treated as fixed constants, and the choice of $g_Y(\cdot)$ may therefore depend on the actual observed Y.

Under the factorization (8), the integral in (7) is rewritten as

$$f(Y) = E_g[(h(Y, S)] \tag{9}$$

and is to be evaluated by Monte Carlo integration using sampling from $g_Y(\cdot)$. Pilot applications, such as those discussed below, reveal the broad applicability and feasibility of these techniques.

We propose to select $g_Y(\cdot)$'s that are based on the following *sequential* factorization of $f(Y, S)$:

$$f(Y,S) = \prod_{t=1}^{T} f(y_t, s_t | I_{t-1}) , \tag{10}$$

where $I_{t-1} = (Y_{t-1}, S_{t-1})$. The relevance of this factorization arises from the fact that DLV models are inherently "sequential", so that assumptions are directly formulated in terms of either one of the two additional factorizations[4]

$$f(y_t, s_t | I_{t-1}) = \begin{cases} f(y_t | s_t, I_{t-1}) \cdot f(s_t | I_{t-1}) , & (11) \\ f(s_t | y_t, I_{t-1}) \cdot f(y_t | I_{t-1}) . & (12) \end{cases}$$

Factorization (11) is naturally associated with DLV models that consist of a latent process for $\{s_t | I_{t-1}\}$ combined with a "measurement" process that determines the observables $\{y_t | s_t, I_{t-1}\}$. A cursory look through the literature seems to suggest that factorization (11) prevails but, in any event, the transformation from (11) to (12) often is straightforward, as the example discussed below illustrates.

Based on the factorizations in (11) and (12), two obvious choices for $g_Y(\cdot)$ and $h(\cdot)$ are

$$g_Y(S) = \prod_{t=1}^{T} f(s_t | y_t, I_{t-1}) , \quad h(Y,S) = \prod_{t=1}^{T} f(y_t | I_{t-1}) , \tag{13}$$

or

$$g_Y(S) = \prod_{t=1}^{T} f(s_t | I_{t-1}) , \qquad h(Y,S) = \prod_{t=1}^{T} f(y_t | s_t, I_{t-1}) . \tag{14}$$

[4] Note that if s does not cause y in the sense of Granger (1969), then

$$f(Y) = \prod_{t=1}^{T} f(y_t | Y_{t-1}) \equiv \prod_{t=1}^{T} f(y_t | I_{t-1}) .$$

The joint density of S given Y then is

$$f(S|Y) = \prod_{t=1}^{T} f(s_t | y_t, I_{t-1})$$

and is immediately available for simulation. This is one of the exceptions we refer to in footnote 2.

In either case, the random draws of S are organized sequentially. At time t, the latent variable s_t is drawn conditionally on the *actual* values of the observables (Y_t in (13) and Y_{t-1} in (14)) and the *simulated* values of S_{t-1}. We note in passing that other factorizations of $f(s_t, y_t|I_{t-1})$ than those just considered may be available and may lead to alternate choices of $g_Y(\cdot)$. An example will be provided below in the context of a simple disequilibrium model.

At this stage of our investigation, it would be premature to attempt ranking the various choice for $g_Y(\cdot)$ in terms of their (potential) Monte Carlo efficiency. A major concern has to be that of minimizing the sampling variance of the residual function $h(\cdot)$. This criterion favors the factorization in (13), since the densities whose product defines $h(\cdot)$ then depend only on the *lagged* values of s_t, while in (14) they also depend on its current value.[5] "Brute force" simulation in the context of the example we discuss below supports our intuition.

There are, however, additional important considerations regarding the choice of $g_Y(\cdot)$. As we discuss further in Section 5 below, "acceleration" or "variance reduction" techniques play a critical role in the practical implementation of Monte Carlo simulation techniques. The (automatic) combination of acceleration techniques in the present context constitutes the major focus of our current investigation. It might be the case that, under some circumstances, more efficient acceleration techniques are available for the factorization in (14) relative to that in (13), in which case our heuristic ranking of the $g_Y(\cdot)$s could easily be reversed.

Since our primary objective here is to demonstrate feasibility, we now discuss a pilot application of the proposed techniques to a problem hitherto considered intractable.

4. An Example

Disequilibrium models have been the focus of much attention over recent years; see the extensive list of references in Quandt (1989). Such models, meant to capture adjustment out of equilibrium, are inherently dynamic (see the discussion in Quandt (1982)), yet truly dynamic versions of these models exhibit complex non-linearities and cannot be estimated by "conventional" techniques. As such, they constitute an ideal test of feasibility for the Monte Carlo simulation techniques we propose to develop. In order to focus attention on the essential elements, we consider only a streamlined version of a dynamic disequilibrium model in which all components other than the relevant lagged variates are deleted, including prices. Extensions to more realistic models are straightforward, conceptually at least.

Let $q_t' = (q_{1t}, q_{2t})$ denote a bivariate random variable whose conditional sampling distribution at time t is Normal, with mean vector and covariance matrix given by

[5] If, in particular, s does not cause y in the sense of Granger (1969), then the sampling variance of $h(Y, S)$ is zero — see footnote 4 — and the choice of $g_Y(\cdot)$ in (13) is optimal.

$$E(q_{it}|q_{t-1}) = \alpha_i\, q_{it-1} \;:\; V(q_t|q_{t-1}) = I\,;\; t:\, 1 \to T\,, \tag{15}$$

where $|\alpha_i| < 1$. The initial conditions are themselves independently normally distributed with zero means and their individual variance in $[1-\alpha_i^2]^{-1}$. Let the sole observable be $y_t = \text{Min}\,(q_{1t}, q_{2t})$. The latent variables are $\lambda_t = \text{Max}\,(q_{1t}, q_{2t})$ and a binary variable δ_t which equals i if $y_t = q_{it}$. Hence, $s_t = (\lambda_t, \delta_t)$.

Both factorizations in (13) and (14) are easily available, based on the standard algebra of the Min operator applied to the bivariate Normal density — see e.g. Quandt (1988). Let $\phi(\cdot)$ denote the standardized Normal density and $\Phi(\cdot)$ denote the corresponding distribution function. Conditionally on (y_{t-1}, λ_{t-1}) and $\delta_{t-1} = 1$, the relevant elements of the factorization in (14) are given by

$$\begin{aligned} f(\lambda_t|I_{t-1}) = {} & \phi(\lambda_t - \alpha_1 y_{t-1})\,\Phi(\lambda_t - \alpha_2\lambda_{t-1}) + \\ & \phi(\lambda_t - \alpha_2\lambda_{t-1})\,\Phi(\lambda_t - \alpha_1 y_{t-1})\,, \end{aligned} \tag{16}$$

$$\begin{aligned} \Pr(\delta_t = 1|\lambda_t, I_{t-1}) = {} & [f(\lambda_t|I_{t-1})]^{-1}\,\phi(\lambda_t - \alpha_2\lambda_{t-1}) \times \\ & \Phi(\lambda_t - \alpha_1 y_{t-1})\,, \end{aligned} \tag{17}$$

$$f(y_t|\lambda_t, \delta_t = 1, I_{t-1}) = \begin{cases} [\Phi(\lambda_t - \alpha_1 y_{t-1})]^{-1}\,\phi(y_t - \alpha_1 y_{t-1}) & \text{if } y_t \le \lambda_t\,, \\ 0 & \text{otherwise,} \end{cases}$$

$$f(y_t|\lambda_t, \delta_t = 2, I_{t-1}) = \begin{cases} [\Phi(\lambda_t - \alpha_2 y_{t-1})]^{-1}\,\phi(y_t - \alpha_2 y_{t-1}) & \text{if } y_t \le \lambda_t\,, \\ 0 & \text{otherwise}\,. \end{cases} \tag{18}$$

Similar expressions are available conditionally on $\delta_{t-1} = 2$ with λ_{t-1} and y_{t-1} being interchanged in formulas (16 ff). The simulation is then based on (16) and (17) and proceeds sequentially. At step t $(t: 1 \to T)$, we first draw λ_t which, according to (16), is the maximum of a bivariate normal distribution. Conditionally on λ_t, we then generate a Bernoulli drawing for δ_t. The function $h(Y, S)$ is the product of T factors of the form given in (18).

In practice, however, this simple algorithm generates an extremely high relative number of "rejections" (i.e. of sequences for which at least one λ_t is lower than the corresponding y_t and, therefore, $h(Y, S)$ is zero), and its efficiency is zero for all practical purposes. This difficulty can be circumvented by constraining the drawings within the "acceptance" region and reweighting them accordingly, an application currently under development.

Conditionally on (y_{t-1}, λ_{t-1}) and $\delta_{t-1} = 1$, the relevant elements of the factorization in (13) are given by

$$\Pr(\delta_t = 1|y_t, I_{t-1}) = [f(y_t|I_{t-1})]^{-1} \cdot \phi(y_t - \alpha_1 y_{t-1}) \times [1 - \Phi(y_t - \alpha_2 \lambda_{t-1})] , \tag{19}$$

$$f(y_t|I_{t-1}) = \phi(y_t - \alpha_1 y_{t-1}) [1 - \Phi(y_t - \alpha_2 \lambda_{t-1})] + \phi(y_t - \alpha_2 \lambda_{t-1}) [1 - \Phi(y_t - \alpha_1 y_{t-1})] , \tag{20}$$

$$f(\lambda_t|\delta_t = 1, y_t, I_{t-1}) = \phi(\lambda_t - \alpha_2 \lambda_{t-1}) [1 - \Phi(y_t - \alpha_2 \lambda_{t-1})]^{-1} , \quad \lambda_t \geq y_t , \tag{21}$$

$$f(\lambda_t|\delta_t = 2, y_t, I_{t-1}) = \phi(\lambda_t - \alpha_1 y_{t-1}) [1 - \Phi(y_t - \alpha_1 y_{t-1})]^{-1} , \quad \lambda_t \geq y_t . \tag{22}$$

Similar expression are available conditionally on $\delta_{t-1} = 2$, with λ_{t-1} and y_{t-1} being interchanged in formulae (19) through (22). The function $h(Y, S)$ is given by the product of T factors of the form given in (20). The simulation is based on (19), (21), and (22) and proceeds sequentially. At step t $(t : 1 \to T)$, we first generate a Bernoulli drawing for δ_t with probabilities as given in (19). Conditionally on δ_t, we then draw λ_t from a truncated Normal distribution given in either (21) or (22), depending on whether δ_t equals one or two.

In the course of the discussion in Section 3, we emphasized the fact that the distributions $g_Y(\cdot)$ are artificial in the sense that they are only meant to approximate the actual distribution of S, given Y. In order to illustrate further this important point, we mention in passing that the sampling process as given in (19), (21), and (22) can be modified at will, at the cost of appropriately redefining the remainder factor in (20). We can, for example, leave (21) and (22) unchanged, but replace (19) by

$$\Pr(\delta_t = 1|y_t, I_{t-1}) = [\phi(y_t - \alpha_1 y_{t-1}) + \phi(y_t - \alpha_2 \lambda_{t-1})]^{-1} \cdot \phi(y_t - \alpha_1 y_{t-1}) . \tag{19'}$$

The function $h(Y, S)$ is redefined accordingly as the product of T factors of the form

$$h(y_t|I_{t-1}) = \begin{cases} \phi(y_1 - \alpha_2 \lambda_{t-1}) \cdot [\phi(y_t - \alpha_1 y_{t-1}) + \phi(y_t - \alpha_2 \lambda_{t-1})] , & \text{if } \delta_t = 1, \\ \phi(y_1 - \alpha_1 y_{t-1}) \cdot [\phi(y_t - \alpha_1 y_{t-1}) + \phi(y_t - \alpha_2 \lambda_{t-1})] , & \text{if } \delta_t = 2. \end{cases} \tag{22'}$$

We have applied the three random number generators just described to a one parameter example when $\alpha_1 = \alpha_2 = \alpha$. Figure 1 describes the outcome of a complete likelihood evaluation for $\alpha = 0.60$ and $T = 50$, based on 10,000 Monte Carlo drawings from the random number generator characterized by Equations (19), (21), and (22). Total computation time is 40 minutes on a Vax 3200 workstation,

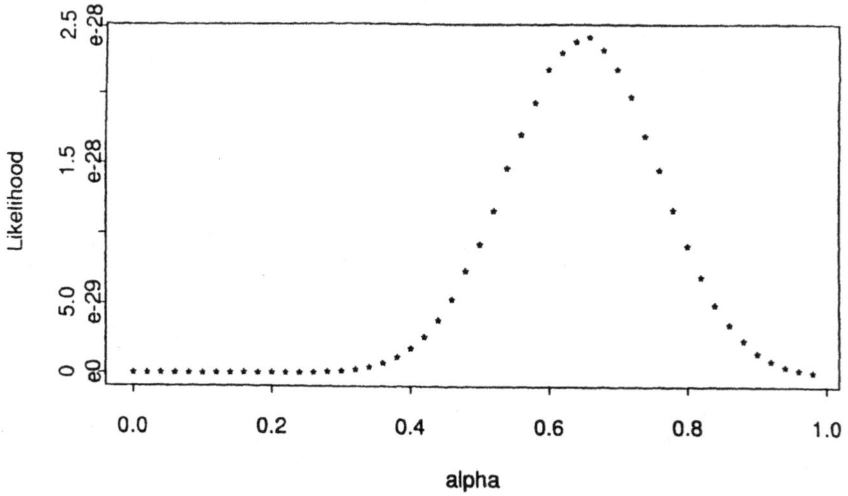

Fig. 1. Dynamic disequilibrium model.

and the Monte Carlo coefficients of variation associated with the individual points in Figure 1 are of the order of 1.5% or less.[6]

The smoothness of the graph in Figure 1 results from using common random numbers for different values of α, which is precisely one of the acceleration techniques described below. For the rest, the application is "brute force", and similar runs using the two other random number generators described above support our intuition that the random number generator associated with factorization (13) is the most efficient (brute force) one by a substantial margin (factors of 5 to 10).

The exceptionally well-behaved shape of the disequilibrium likelihood function in Figure 1 will be a surprise to anyone familiar with the pathologies of disequilibrium models and results from the implicit constraints imposed on the coefficients of the model (known variances and equal slopes). Our maximally streamlined example can be extended to richer two-equation models by the introduction of additional regressors in (15) and by the relaxation of the parameter restrictions. An additional price adjustment equation may be added without problem — an important addition if price movements are thought to be useful indicators of disequilibrium. We expect

[6] Whether or not such precision "suffices" depends very much on the final objective of such a likelihood evaluation. It would probably prove quite sufficient for a Bayesian statistician who would multiply the likelihood function by a prior density and rescale the product to obtain the posterior density of α. If one's objective is instead that of computing a ML estimator, the matter becomes one of evaluating the relative contributions of the Monte Carlo uncertainty regarding the exact localization of the estimator itself and its (classical) sampling variance. Note that an overnight run of computation would reduce the Monte Carlo coefficients of variation by a factor of 4 and, more importantly, that we expect to achieve far greater efficiency gains with the acceleration techniques described in Section 5.

to gain an understanding of the pathologies of dynamic disequilibrium models from these extended models and hence of their impact on the above approach. This is not the least exciting prospect offered by the availability of techniques whereby the likelihood functions of models previously considered intractable can now be explored.

5. Acceleration Techniques

A common criticism of Monte Carlo techniques is their potential inefficiency. Fortunately, there now exists a number of conceptually simple and powerful variance reduction techniques that can result in dramatic computational gains (see Hendry (1984) and Geweke (1988a, 1988b)).

The main variance reduction techniques, which are briefly discussed below, are antithetic variates, control variates, and common random numbers. Within the context of DLV models, only the latter two are in value, especially when used in conjunction with one another. The thrust of our research in this area is the development of control variates. In the subsequent discussion, ψ denotes a quantity to be simulated and ψ^* its brute force Monte Carlo estimate. Alternative estimates are ψ^-, ψ^+, ψ' and $\bar{\psi}$.

1. *Antithetic Variates (AVs)*. The object of AVs is to construct a matching pair of estimates ψ^- and ψ^+ that are highly negatively correlated and hence able to offset each other's variability. A pooled estimate $\bar{\psi} = \frac{1}{2}(\psi^- + \psi^+)$ is evaluated that will have lower variance than either of the individual estimates. The most common practice for symmetric distributions is that of employing a sequence $\{\varepsilon_t\}$ of random numbers for ψ^+ and the matching sequence $\{-\varepsilon_t\}$ for ψ^-. The increase in the cost of computation is clearly minimal and, as reported in Geweke (1988a), the procedure may generate considerable efficiency gains under the appropriate circumstances.

It is, however, unlikely that AV procedures will prove of much use in our context (see Hendry and Trivedi (1990)). If, for example, we compute the OLS estimate of α in an equation of the form

$$y_t = \alpha y_{t-1} + \varepsilon_t\,, \quad y_0 = 0\,, \tag{23}$$

then replacing $\{\varepsilon_t\}$ by $\{-\varepsilon_t\}$ leaves the OLS estimate unchanged, and AVs become essentially useless. The picture is somewhat more complicated in the context of a disequilibrium example, although numerical experimentation indicates that AV pairs of estimates of the likelihood function are essentially uncorrelated.

2. *Control Variates (CVs)*. This technique amounts to reusing the random numbers $\{\varepsilon_t\}$ to construct an estimate ψ' of known expectation that is positively correlated with ψ^*. A pooled estimate is then given by $\bar{\psi} = (\psi^* - \psi') + E(\psi')$ and, under suitable conditions, its variance may be substantially lower than that of ψ^*.

The choice of ψ' is often *ad hoc*, and many such CV estimates can be constructed for any given application — which at first sight seems to defeat the very purpose of our investigation. However, there exist classes of CV estimates that are easily implementable in the context of DLV models and seem potentially efficient based on two complementary lines of attack.

First, Davidson and MacKinnon (1989) suggest ways of obtaining CV estimates using various functions of the random numbers and also suggest that the weight to be given to ψ' in the definition of the pooled estimate be determined by means of regression after the simulation has been completed. Second, DLV models generally have the property that their likelihood functions can be analytically evaluated over apropriate subsets of the parameter space. Hence, let the parameter θ be partitioned into (γ, β) and assume that the likelihood function is known at (γ_0, β). Likelihood estimates at (γ_0, β) may then be used as CVs for likelihood estimates at "neighboring" values of (γ, β) and additional efficiency gain could conceivably be achieved by means of a series expansion around (γ_0, β). In the context of our disequilibrium model, the exact likelihood at $\alpha = 0$ is given by

$$L(0; y) = \prod_{t=1}^{T} 2\phi(y_t) \cdot [1 - \Phi(y_t)] \,. \tag{24}$$

Note, however, that for $\alpha = 0$, the $h(\cdot)$ functions associated with equations (20) and (22') do not depend on the latent variables and cannot therefore serve as CVs.[7] This approach can be implemented in the context of the factorization (16)–(18). To illustrate at a simpler level how the proposed technique operates, consider the following one parameter DLV logit model

$$s_t = \alpha s_{t-1} + \varepsilon_t \,, \quad \varepsilon_t \sim N(0, 1) \,, \tag{25}$$

$$\Pr(y_t = 1 | s_t, I_{t-1}) = [1 + \exp(-s_t)]^{-1} = p(s_t) \,. \tag{26}$$

The obvious random generator for this example is that associated with equation (25). The corresponding $h(\cdot)$ function is given by

$$h(Y, S) = \prod_{t \in T_1} [p(s_t)] \cdot \prod_{t \in T_2} [1 - p(s_t)] \,, \tag{27}$$

where $T_1 = \{t; y_t = 1\}$ and $T_2 = \{t; y_t = 0\}$. An obvious CV is given by

$$h'(\cdot) = \prod_{t \in T_1} [p(\varepsilon_t)] \cdot \prod_{t \in T_2} [1 - p(\varepsilon_t)] \,. \tag{28}$$

Its expectation $(\frac{1}{2})^T$. Numerical investigation indicates that not only is this CV efficient in the immediate neighborhood of $\alpha = 0$, but that its range of usefulness

[7] This serves to illustrate an earlier point that the ranking between different g_Y's may be affected by the availability of acceleration techniques.

extends up to values of α less than .5 in absolute value. Refinements of this technique, in which CVs could depend on α, deserve investigation.

3. *Common Random Numbers (CRNs).* If interest focuses on a statistic across the parameter space, it is natural to reuse the same random numbers at different points θ_i. In doing so, we introduce substantial positive autocorrelation across evaluations at different θ_i and, hence, reduce the variance of the difference between the statistic at neighboring values of θ. In our context, this procedure smooths the estimators of the likelihood functions and, judging by Figure 1, the results are visually quite impressive. Note, however, that this smoothness induces an artificial sense of precision since the location of the function remains open to uncertainty. Insofar as the likelihood function needs only to be known up to a multiplicative constant, this uncertainty may not prove critical.

Even more promising is the combination of CV and CRN techniques. In fact, the CV technique discussed above combines both since it relies upon the use of CRNs across neighboring values of θ. Whatever efficiency gains can be achieved at specific points in the parameter space by means of CV techniques spill over to neighboring values of θ when CRNs are used.

6. Exogeneity

Applications of MSM techniques have largely been confined to (essentially)[8] static models, so that the treatment of "exogenous" variables raises no additional problems. In a dynamic context, however, simulation under *fixed* exogenous variables (or "regressors") requires the assumption that the latter be *strongly* exogenous, in the sense of Engle *et al.* (1983). This requires that, in addition to the usual (weak) exogeneity assumptions, the *latent* variables cannot feedback onto the exogenous process, in the sense of Granger (1969).

Let us be more specific on this important issue. In line with the notation in Section 3, let x_t denote a vector of (weakly) exogenous variables. The notation X_{t-1} represents the matrix $(x_1 \dots x_{t-1})$. Let $W_{t-1} = (S_{t-1}, X_{t-1}, Y_{t-1})$ regroup the lagged values of all relevant variables. The symbol $\perp$ denotes independence in probability (conditional independence when followed by the conditioning operator $|$). The result we need in order to validate likelihood simulation is one that follows from the analysis in Engle *et al.* (1983) and that is summarized in the form of the following lemma:
Lemma: If

$$x_t \perp S_{t-1} | X_{t-1}, Y_{t-1} \ , \tag{29}$$

then the likelihood function of the observables (Y, X), when X is a matrix of weakly exogenous variables, is given by

[8] Occasionally allowing for autocorrelation in the disturbance terms.

$$L(\theta; Y, X) = \int .. \int \Big[\prod_{t=1}^{T} f(y_t, s_t | x_t, W_{t-1}, \theta)\Big]\, ds_1, .. ds_T\,. \tag{30}$$

Proof: Immediate, since (29) implies that

$$f(x_t | W_{t-1}, \cdot) \equiv f(x_t | Y_{t-1}, \cdot)\,.$$

The latter density does not depend on the latent variables and factors out of the integral in (30). Q.E.D.

Compare formula (30) with formulae (7) and (10). It follows that under condition (29), the exogenous variables are kept fixed at their *observed values* in the evaluation of (30). If, instead, assumption (29) does not hold, then the exogenous process also belongs to the integral in (30).

We note in passing that an exogeneity analysis also has obvious implications for the use of MSM techniques in the context of dynamic models. If both the latent and the observable endogenous variables are subject to simulation, then assumption (29) has to be replaced by the stronger condition

$$x_t \perp (S_{t-1}, Y_{t-1}) \,|\, X_{t-1}\,. \tag{31}$$

7. Conclusion

The simulation-based methods discussed above for evaluating the likelihood function of dynamic models involving latent variables offer considerable promise in terms of rendering tractable previously (almost) intractable problems, especially those concerning disequilibrium (min-condition) models. We have focused on likelihood evaluation because of its central role in estimation and inference in both classical and Bayesian approaches. As experience accumulates with alternative simulation devices and conditional factorizations, we anticipate the development of operational methods for modelling small dynamic systems with both latent and unmodelled (weakly exogenous) variables that allow an appropriate treatment of practical model selection problems despite the complexity of the joint density function for the observables and latent variables. Further extensions to non-stationary cases (e.g. cointegrated sub-systems with some unit root processes) should also prove feasible.

References

Aigner, D.J., C. Hsiao, A. Kapteyn and T. Wansbeek: 1983, 'Latent variables models in econometrics', Ch. 23, in *The Handbook of Econometrics*, edited by Z. Griliches and M.D. Intriligator, Amsterdam: North Holland.

Andrews, D.W.K.: 1989, 'An empirical process Central Limit theorem for dependent non-identically distributed random variables', Cowles Foundation Discussion Paper, no. 907, Yale University.

Davidson, R. and J.G. MacKinnon: 1989, 'A general procedure for using control and antithetic variables in Monte Carlo experiments', GREQE Discussion Paper, no. 89A09, University of Aix-Marseille.

Duffie, D. and K.J. Singleton: 1989, 'Simulated moments estimation of Markov models of asset prices', Stanford University mimeo.

Engle, R.F., D.F. Hendry, and J.-F. Richard: 1983, 'Exogeneity', *Econometrica* **51**, 277–304.

Geweke, J.: 1988a, 'Antithetic acceleration of Monte Carlo integration in Bayesian inference", *Journal of Econometrica* **38**, 73–90.

Geweke, J.: 1988b, 'Acceleration methods for Monte Carlo integration in Bayesian inference', in *Proceedings of the 20th Symposium on the Interface: Computing Science and Statistics*, forthcoming.

Geman, S. and D. Geman: 1984, 'Stochastic Relaxation, Gibbs Distributions and the Bayesian Restoration of Images', *IEEE Transactions on Pattern Analysis and Machine Intelligence* **6**, 721–741.

Granger, C.W.J.: 1969, 'Investigating causal relations by econometric models and cross-spectral methods', *Econometrics* **37**, 424–438.

Hajivassiliou, V.A. and D.L. McFadden: 1989, 'The method of simulated scores for the estimation of LDV models with an application to the problem of external debt crisis', Yale University, mimeo.

Hajivassiliou, V.A.: 1989, 'Macroeconomic shocks in an aggregate disequilibrium model', Yale University, mimeo.

Hansen, L.P.: 1982, 'Large sample properties of generalized method of moments estimates', *Econometrica* **50**, 1027–1054.

Heckman, J.J.: 1981, 'Statistical models for discrete panel data', Ch. 3, in *Structural Analysis of Discrete Data with Economic Applications*, edited by C.F. Manski and D. McFadden, Cambridge: MIT Press.

Heckman, J.J. and T.F. McCurdy: 1980, 'A life cycle model of female labor supply', *The Review of Economic Studies* **47**, 47–74.

Hendry, D.F.: 1976, 'The structure of simultaneous equations estimators', *Journal of Econometrics* **4**, 51–88.

Hendry, D.F.: 1984, 'Monte Carlo experimentation in econometrics', Ch. 16 in *The Handbook of Econometrics*, edited by Z. Griliches and M.D. Intriligator, Amsterdam: North Holland.

Hendry, D.F. and P.K. Trivedi: 1972, 'Maximum Likelihood Estimation of Difference Equations with Moving Average Errors: A Simulation Study', *Review of Economic Studies* **39**, 117–145.

Laffont, J.J. and A. Monfort: 1979, 'Disequilibrium econometrics in dynamic models', *Journal of Econometrics* **11**, 353–61.

Lerman, S. and C. Manski: 1981, 'On the use of simulated frequencies to approximate choice probabilities', in *Structural Analysis of Discrete Data with Econometric Applications*, edited by C. Manski and D. McFadden, Cambridge: MIT Press.

Maddala, G.S.: 1983, 'Disequilibrium, self-selection and switching models', Ch. 28, in *The Handbook of Econometrics*, edited by Z. Griliches and M.D. Intriligator, Amsterdam: North Holland.

McFadden, D.: 1989, 'A method of simulated moments for estimation of discrete response models without numerical integration', *Econometrica* **57**, 995–1026.

Pakes, A.: 1986, 'Patents as options: Some estimates of the value of holding European patent stocks', *Econometrica* **54**, 755–784.

Pakes, A. and D. Pollard: 1989, 'Simulation and the asymptotics of optimization estimation', *Econometrica* **57**, 1027–1058.

Quandt, R.E.: 1982, 'Econometric disequilibrium models', *Econometric Review 1*, 1–96 (with comments by D.F. Hendry, A. Monfort and J.-F. Richard).

Quandt, R.E.: 1988, *The Econometrics of Disequilibrium*, New York: Basil Blackwell.

Quandt, R.E.: 1989, 'Bibliography of quantity rationing and disequilibrium models', Princeton University, Department of Economics.

GLOBAL OPTIMIZATION OF STATISTICAL FUNCTIONS: PRELIMINARY RESULTS[1,2]

WILLIAM L. GOFFE
Southern Methodist University

GARY D. FERRIER
Southern Methodist University

and

JOHN ROGERS
University of North Texas

I. Introduction

Many econometric methods, such as nonlinear least squares, the generalized method of moments, and the maximum likelihood method, rely upon optimization to estimate model parameters. Not only do these methods lay the theoretical foundation for many estimators, they are often used to numerically estimate models. Unfortunately, this may be difficult because these algorithms are sometimes touchy and may fail. For the maximum likelihood method, Cramer (1986, p. 77) lists a number of "unpleasant possibilities": the algorithm may not converge in a reasonable number of steps, it may diverge toward ridiculous values, or even loop through the same point time and again. Also, the algorithm may have difficulty with ridges and plateaus. When faced with such difficulties, the researcher is often reduced to trying different starting values (Cramer, 1986, p. 72 and Finch et al., 1989). Finally, even if the algorithm converges, there is no assurance that it will have converged to a global, rather than a local optimum since conventional algorithms cannot distinguish between them. In sum, there is a poor match between the power of these methods and the algorithms used to implement them.

A new and very different optimization algorithm, simulated annealing, potentially improves this match. It explores the function's entire surface and slowly tries to optimize the function while moving both uphill and downhill.

[1] We would like to thank, without implicating, T. Randolph Beard and David Levy for comments on a previous draft. For computational help, we would like to thank Art Belmonte and Ava Horvath of IMSL. Sarah Barron, Robert Harkness, John Respess and Spiros Vellas of the University of Texas Center for High Performance Computing also provided invaluable help.

[2] The computational work for this paper was performed at the University of Texas Center for High Performance Computing.

Hans M. Amman et al. (eds), Computational Economics and Econometrics, 19-32.

Thus, it is largely independent of the starting values, often a critical input in conventional algorithms. Further, it can escape from local optima and go on to find the global optimum by the uphill and downhill moves. Simulated annealing also makes less stringent assumptions regarding the function than conventional algorithms, so it can more easily deal with ridges and plateaus. Finally, it can optimize functions that do not exist for some parameter values.

This paper compares the Corana et al. (1987) implementation of simulated annealing to conventional algorithms on an econometric model. Thus, these results may help guide other researchers. The model is an example from the literature with multiple minima (Judge et al., 1985, pp. 956-7). The three multivariable optimization algorithms in the IMSL library represent conventional algorithms. Since they are sensitive to starting values, 100 different sets of them are used.

Essentially, simulated annealing provides more features at the cost of an increase in runtime. As shown in Appendix A, computer power is increasing rapidly and the resources needed for simulated annealing will soon, if not already, be available to most researchers for even the most difficult problems.

This paper is organized in the following fashion. Section II describes simulated annealing in detail and compares it to conventional algorithms, while section III compares its performance on the test problem to the IMSL algorithms. A conclusion summarizes the strengths and weaknesses of simulated annealing.

II. The Algorithms

A. Conventional Algorithms and Their Limitations

Conventional optimization algorithms try to move uphill (maximization is assumed here) in an iterative manner. In very general terms, starting from a point, they determine the best direction and step length to head uphill. After moving there, this process is repeated until some stopping criteria is achieved. Statistical packages such as SAS, TSP and RATS use these algorithms. General reviews on them can be found in either Judge et al. (1985) or Press et al. (1986), and a more rigorous analysis is in Dennis and Schnabel (1983). Their operation is similar to a blind man walking up a hill whose only knowledge of the hill comes from what passes under his feet. If the hill is predictable in some fashion, he will reach the top, but it is easy to imagine confounding terrain. In contrast, simulated annealing builds up a view of the entire hill (or range of hills) and then maximizes the function. Most conventional algorithms make the function predictable by assuming the function is approximately quadratic. Implicit here is another assumption common to all conventional algorithms: the function has one optimum, so that any local optimum is also the global optimum.

Unfortunately, statistical functions may not be approximately quadratic. Further, while Kendall and Stuart (1978, p. 44) show asymptotically that likelihood functions have one optimum, the small sample properties are unknown. Taken together, it is easy to see why conventional algorithms may have difficulty finding the maximum.

B. Simulated Annealing for Continuous Variable Problems

Simulated annealing's roots are in thermodynamics, where one studies a system's thermal energy. A description of the cooling of molten metal motivates this new optimization technique. After slow cooling (annealing), the metal arrives at a low energy state. Inherent random fluctuations in energy allows the annealing system to escape local energy minima to achieve the global minimum. But if cooled very quickly (or "quenched"), it might not escape local energy minima and when fully cooled it may contain more energy than annealed metal. Simulated annealing attempts to minimize some analogue of energy in a manner similar to annealing to find the global minmum. Details can be found in Press et al. (1986, pp. 326-34).

Early simulated annealing algorithms considered combinatorial systems, where the system's state depends on the configuration of variables. Perhaps the best known is the traveling salesman problem, in which one tries to find the minimum trip distance between a number of cities. Combinatorial simulated annealing has been used successfully in computer and circuit design (Kirkpatrick et al., 1983, and Wong et al., 1988), pollution control (Derwent, 1988), a special case of 0-1 programming (Drexl, 1988), neural networks, reconstruction of pollycrystalline structures (Telly et al., 1987) and image processing (Carnevali et al., 1985).

Global optimization algorithms with different bases have been introduced in recent years. They include adaptive random search (Pronzato et al., 1984), genetic algorithms (Golderg, 1989) and a method using stochastic differential equations (Aluffi-Pentini et al., 1988). Both Vanderbilt et al. (1984) and Bohachevsky et al. (1986) have modified simulated annealing for continuous variable problems. However, the Corana et al. (1987) implementation of simulated annealing for continuous variable problems appears to offer the best combination of ease of use and robustness, so it is used in this study. While a complete description can be found there, a summary follows and the algorithm in pseudo-computer code is found in Appendix B. The essential starting parameters are T_0, the initial temperature; X, the starting vector of parameters that will maximize the model; and V, the step length for X. Note that X and V are both vectors of length n, the number of parameters of the model upper case refers to vectors and lower case to scalars (with the exception of temperature, T).

A function evaluation is made at the starting point X and its value f is recorded. Next, a new X, X′, is chosen by varying element i of X,

$$x_i' = x_i + r \cdot v_i \ , \tag{1}$$

where r is a uniformly distributed random number from [-1, 1] and v_i is element i of V. The function value f′ is then computed. If f′ is greater than f, X′ is accepted, X is set to X′ and the algorithm moves uphill. If this is the largest f, it and X are recorded since this is the best current value of the optimum. If f′ is less than or equal to f, the Metropolis criteria decides on acceptance (thermodynamics motivates this criteria). The value

$$p = e^{\left(\frac{f'-f}{T}\right)} \tag{2}$$

is computed and compared to p′, a uniformly distributed random number from [0, 1]. If p is greater than p′, the new point is accepted and X is updated with X′ and the algorithm moves downhill. Otherwise, X′ is rejected. Two things decrease the probability of a downhill move: reductions in temperature and increases in the size of the move itself. After N_S such steps through all elements of X, the step length vector V is adjusted so that 50% of all moves are accepted. The goal here is to sample the function widely. If a greater percentage of points are accepted for x_i, then the relevant element of V is enlarged. For the same temperature, this increases the number of rejections and decreases the percentage of rejections. After N_T times through the above loops, the temperature is reduced. The new T is given by

$$T' = r_T \cdot T \ , \tag{3}$$

where r_T is between 0 and 1. A lower temperature makes a given downhill move less likely, so the number of rejections increase and the step lengths decline. In addition, the first point tried at the new temperature is the current optimum. The smaller steps and starting at the optimum focuses attention on the most promising area.

The algorithm ends by comparing the last N_ε values of the largest function values from the end of each temperature reduction with the most recent one and the optimum function value. If all these differences are less than ε, the algorithm terminates. This criteria helps ensure that the global maximum is reached.

Note that simulated annealing first builds up a rough view of the surface with large step lengths. As the temperature falls and the step length declines, it slowly focuses on the most promising area. All the while, the algorithm can escape from local maxima through downhill moves. Eventually, the algorithm should converge to the function's global maximum.

Simulated annealing has several potential advantages over conventional algorithms. First, it can escape from local maxima. In thermodynamic terms, while conventional algorithms quench by simply heading up the current hill without regard to others, simulated annealing moves both uphill and downhill. Also, the function need not be approximately quadratic. In fact, it need not even be differentiable (Corana et al. (1987) successfully demonstrate simulated annealing on a parabolic function punctured with holes). Another benefit is that the step length gives the researcher valuable information about the function. If an element of it is very large, the function is very flat in that parameter. Since it is determined by function evaluations at many points, it is a better overall measure of flatness than a gradient evaluation at a single point. Finally, simulated annealing can identify corner solutions because it can "snuggle" up to a corner for functions that don't exist in some region. The most important advantage of simulated annealing is that it can maximize functions that are difficult or impossible otherwise. The sole drawback to simulated annealing is the computational power required since the function is evaluated repeatedly. However, as described in Appendix A, major improvements are occurring in high performance computing. To put it briefly, more than half the power of the first Cray supercomputer can be put on a desktop. Further, this trend shows no sign of slowing. Thus, simulated annealing is an attractive option for difficult functions.

III. Simulated Annealing Compared To Conventional Algorithms

A. Comparison Framework

The previous section demonstrated that simulated annealing has promise for optimizing statistical functions. To see if this promise holds, simulated annealing was compared to three conventional algorithms. The purpose here is to see if simulated annealing works on this problem.

All three multivariate optimization algorithms from the IMSL Math/Library Edition 10 were chosen for comparison with simulated annealing. This library was chosen because of its quality and availability. The routines are UMPOL, a simplex algorithm, UMCGF, a conjugate gradient algorithm with numerical derivatives, and UMINF, a quasi-Newton algorithm with numerical derivatives. The numerical derivative versions of UMCGF and UMINF were chosen because computing the analytical derivatives is sometimes very complex. Note that the simplex algorithm does not use derivatives. Since conventional algorithms are sensitive to starting values, each test with them used one hundred different randomly selected starting values. This number surely gives the algorithm a good chance of finding the optimum. Plus, the typical researcher is unlikely to use more.

B. Test Function

Judge et al. (1985, pp. 956-7) provide the test function. This nonlinear least squares problem is written in minimization form and has two local minima. It has two parameters and 20 observations. While small and artificial, it serves as a convenient test from the literature for optimization algorithms because of the two local minima. As will be seen, conventional algorithms cannot by themselves distinguish between the two, while simulated annealing has no difficulty. Note that it is optimized in minimization form.

C. Computing Environment

All computing was done at the University of Texas Center for High Performance Computing in Austin on a Cray X-MP/24 (it was easiest to do all computing at one site even though this performance was not always needed). All programs were written in Fortran 77 with some Fortran 90 (which includes many features that simplify writing programs with arrays) extensions using the CFT77 compiler, versions 2.0 and 3.0. A 64 bit representation was used for all floating point numbers. While single precision on a Cray, this is double precision in the Vax or IBM worlds.

The Cray presented one difficulty. When a floating point error occurs on it, the program terminates. This is different from the IBM mainframe world where a software correction often takes place and execution continues. Thus, it would appear that results generated on a Cray may not carry over to other computers. However, the Cray handles an unusually wide range of numbers, from 10^{2466} to 10^{-2466}. Thus, floating point errors are only likely to occur if the algorithm is out of control. In these situations, a computer with a fixup for floating point errors is unlikely to do better. Thus, the Cray's lack of a fixup for floating point errors should not limit the lessons learned here.

D. Results with the Test Function

Table 1 shows the results of all algorithms for the Judge model. All algorithms were run 100 times with different starting values for the model parameters (the same ones were used by all algorithms). A uniformly distributed random number

TABLE 1
Comparison of all Algorithms on Judge's Function

Bounds on Starting Values: $-100 < \Theta_i < 100$, i=1, 2
Number of Runs: 100

Algorithm	UMPOL (simplex)	UMCGF (conj. gradient)	UMINF (quasi-Newton)	SA (sim. ann.)
Solutions[a]				
@20.482	40	48	48	0
@16.082	60	52	52[b]	100
Mean Number of Function Evaluations	152.54	c	31.25	845.241[d]
Average Execution Time (seconds)	0.019678	0.004083	0.006829	16.328

[a] Solutions are categorized by the minimum at which they terminate. The minimum with the value of 16.082 is the global minimum.
[b] The algorithm twice reports, "The last global step failed to locate a lower point than the current X value. The current X may be an approximate local minimizer and no more accuracy is possible or the step tolerance may be too large where STEPTL = 3.696522E-10." All other terminations from all algorithms are reported as normal.
[c] UMCGF does not report the number of function evaluations.
[d] This number is much smaller when less conservative valuesfor the parameters for simulated annealing were employed.

TABLE 2
Simulated Annealing on the Judge Function: Sample Output from Selected Temperatures

Initial Parameters for the Algorithm							
Initial Parameters for the Likelihood Function: -13.722, -77.314 Resulting Function Value: 1.228806E+08							
T_0:	5,000,000	ε:	1.0E-8	N_S:	20	C_i:	2.0 (i = 1,2)
δ:	0.85	N_ε:	4	N_T	100	V_i:	100.0 (i = 1,2)

Temperature	4,250,000.00
Current Optimal Parameters	5.8319, -2.9451
Current Optimal Function Value	523.7317281388
Downhill, Accepted Uphill, & Rejected Moves	1011, 1016, 1973
Stepsize	1118.4, 166.51
Temperature	18.37500
Current Optimal Parameters	0.87094, 1.2255
Current Optimal Function Value	16.08428345998
Downhill, Accepted Uphill, & Rejected Moves	981, 1001, 2018
Stepsize	1.6176, 1.7525
Temperature	0.7944483E-4
Current Optimal Parameters	0.86479, 1.2357
Current Optimal Function Value	16.08173013579
Downhill, Accepted Uphill, & Rejected Moves	992, 1009, 1999
Stepsize	0.50138E-2, 0.24662E-2
Temperature	0.2414776E-08
Current Optimal Parameters	0.86479, 1.2357
Current Optimal Function Value	16.08173013296
Downhill, Accepted Uphill, & Rejected Moves	1012, 994, 1994
Stepsize	0.23341E-4, 0.22194E-4
Return from Simulated Annealing	Normal
Solution for Parameters	0.8644876261063, 1.235748322078
Optimal Function Value	16.08173013296
Final Stepsize	0.23341E-4, 0.22194E-4
Final Temperature	2.4147761862962E-9
Number of Accepted Moves	434,015
Number of Function Evaluations	868,001
Execution Time	16.0518 seconds

generator selected them from -100 to 100 to simulate the researcher's uncertainty about the best starting value. For the IMSL routines, their suggestions for algorithm inputs were used where given, except the maximum number of function evaluations was set to a very large value to avoid early termination. Otherwise, neutral values were chosen. The inputs for simulated annealing are in Table 2.

The top of Table 1 lists the results from the algorithms. In all cases they converged to one of the two local minima (the global minimum has a value of 16.082, while a nearby local minimum has a value of 20.482). The conventional algorithms were almost evenly split between them, while simulated annealing always found the global minimum.

The remaining part of Table 1 lists the mean number of function evaluations and mean execution time for the 100 runs. As expected, the simplex algorithm takes longer than the other conventional algorithms (Press et al., 1986, p. 289). The longer time for simulated annealing is partly due to using Corana et al.'s (1987) very conservative suggestions for parameters (their paper deals with very difficult functions). Even with less conservative values (in particular, r_T and N_T can be reduced), simulated annealing requires more computational resources than conventional algorithms.

Table 2 shows the results of one simulated annealing run. It first lists the values for the input parameters. With the exception of T_0, V and ε, all have values suggested by Corana et al. (1987). The initial temperature, T_0, was chosen so that the step length was approximately 100 in both parameters (different temperatures were tried until this was achieved). This size ensures that the function is well searched. Correspondingly, the initial value of V is 100 (since the algorithm adjusts V, this initial parameter is not very important). The parameter ε was chosen to ensure that the solution has converged to the global minimum. It is slightly larger than the final tolerances used in the quasi-Newton and simplex algorithms (the conjugate gradient documentation is silent on this matter).

The table next lists some intermediate output as the temperature falls. For each of the selected temperatures (from more than 216), the current best parameter values along with the resulting function value are shown. Also listed are the number of downhill and uphill moves. Downhill evaluations are always accepted, while uphill moves are accepted according to the Metropolis criteria. These results indicate that the step length is chosen correctly since about half of all functions evaluations are accepted (recall the step length adjusts to ensure this). The number of function evaluations at each temperature is $N_S \cdot N_T \cdot n$, or 4,000. Finally, the step length for both parameters is shown.

As the temperature falls, one can watch the algorithm closing in on the global minimum. The falling temperature along with the Metropolis criteria causes fewer uphill moves to be accepted. This rise in rejections in turn shrinks the step length. In conjunction with the algorithm starting out at the existing optimum with each new temperature, the smaller step length focuses the algorithm on the most promising area. The final part of the table shows the results produced by a successful termination of the algorithm.

V. Conclusion

This paper tests a recently developed algorithm, simulated annealing, on a statistical model that requires optimization for parameter estimation. Compared to the three conventional optimization algorithms from the IMSL library, simulated annealing is shown to be superior.

Simulated annealing correctly differentiated between the global and local optima, while the conventional algorithms did not. This shows that simulated annealing has considerable promise for optimizing statistical functions.

References

Aluffi-Pentini, Filippo, Valerio Parisi and Francesco Zirilli. "A Global Optimization Algorithm Using Stochastic Differential Equations." *ACM Transactions on Mathematical Software* 14 (1988): 344-65.

Bell, Gordon. "The Future of High Performance Computers in Science and Engineering." *Communications of the ACM* 32 (1989): 1091-101.

Bohacnebsky, Ihor O., Mark E. Johnson and Myron L. Stein. "Generalized Simulated Annealing for Function Optimization." *Technometrics* 28 (1986): 209-17.

Carnevali, P., L. Coletti and S. Patarnello. "Image Processing by Simulated Annealing." *IBM Journal of Research and Development* 29 (1985) 569-79.

Corana, A., M. Marchesi, C. Martini and S. Ridella. "Minimizing Multimodal Functions of Continuous Variables with the Simulated Annealing' Algorithm." *ACM Transactions on Mathematical Software* 13 (1987): 262-80.

Cramer, J. S. *Econometric Applications of Maximum Likelihood Methods*. New York: Cambridge University Press, 1986.

Dennis, J. E. and Robert B. Schnabel. *Numerical Methods for Unconstrained Optimization and Nonlinear Equations*. Englewood Cliffs, NJ: Prentice-Hall, 1983.

Derwent, Dick. "A Better Way to Control Pollution." *Nature* 331 (1988): 575-78.

Dongarra, Jack J. "Performance of Various Computers Using Standard Linear Equations Software in a Fortran Environment." Mathematics and Computer Science Division, Argonne National Laboratory, Argonne, IL 1988.

Drexl, A. "A Simulated Annealing Approach to the Multiconstraint Zero-One Knapsack Problem." *Computing* 40 (1988): 1-8.

Finch, Stephen J., Nancy R. Mendell and Henry C. Thode, Jr. "Probabilistic Measures of Adequacy of a Numerical Search for a Global Maximum." *Journal of the American Statistical Association* 84 (1989): 1020-3.

Ferrier, Gary. "A Comparison of Alternative Analytic Techniques Available for the Study of Production Economies and Efficiencies: An Application To the U.S. Banking Industry." Diss.

The University of North Carolina at Chapel Hill, 1988.

Goldberg, David E. *Genetic Algorithms in Search, Optimization and Machine Learning*. Reading MA: Addison-Wesley, 1989.

Judge, George G., W. E. Griffiths, R. Carter Hill, Helmut Lutkepohl and Tsoung-Chao Lee. *The Theory and Practice of Econometrics*. 2nd ed. New York: John Wiley and Sons, 1985.

Kendall, Maurice and Alan Stuart. *The Advanced Theory of Statistics*. 4th ed. New York: Macmillan, 1978.

Kirkpatrick, S., C. D. Gelatt Jr. and M. P. Vecchi. "Optimization by Simulated Annealing." *Science* 220 (1983): 671-80.

MIPS Staff. "486 Times Two, Intel's iSBC 486/125 and IBM's 486/25 Power Platform Give CISC's Answer to RISC." *MIPS* October 1989: 36-4.

Pronzato, Luc, Eric Walter, Alain Venot and Jean-Francois Lebruchec. "A General-Purpose Optimizer: Implementation and Applications." *Mathematics and Computers in Simulation* 24 (1984): 412-22.

Press, William H., Brian Flannery, Saul A. Teukolsky and William T. Vetterling. *Numerical Recipes, The Art of Scientific Computing*. New York: Cambridge UP, 1986.

RISC System/6000. White Plains, NY: IBM US Marketing and Services, 1990.

Telly, H., Th. M. Liebling and A. Mocellin. "Reconstruction of Polycrystalline Structures: a New Application of Combinatorial Optimization." *Computing* 38 (1987): 1-11.

Vanderbilt, David and Seven G. Louie. "A Monte Carlo Simulated Annealing Approach to Optimization over Continuous Variables." *Journal of Computational Physics* 56 (1984): 259-71.

Wong, D. F., H. W. Leong and C. L. Liu. *Simulated Annealing for VLSI Design*. Boston: Kluwer Academic Publishers, 1988.

Appendix A: Computer Performance

The most useful widely used benchmark of computer performance for econometric work is probably the Linpack benchmark. In it, a system of 100 linear equations are solved and the number of floating point operations per second are counted (mflops for a million of them). This is an appropriate benchmark since it involves floating point operations on arrays.

Table A1 (from Dongarra) shows the number of floating point operations per second on the Linpack benchmark for some existing computers. The Cray X-MP is Cray's best selling machine and all work for this paper was done on one. It has recently been succeeded by the Y-MP, which has about twice the performance. The Cray 1S was Cray's first widely used machine and is still used as a reference. The IBM 3090 is currently IBM's most powerful computer and as an option it can be equipped with a vector processor for greater performance. The IBM 3081 was the top of IBM's mainframe line in the early 1980's, and the IBM PC is well known.

TABLE A1

COMPUTER	MFLOPS*
Cray X-MP	39.0000
Cray 1S	12.0000
IBM 3090/180E (w/ vector processor)	13.0000
IBM 3090 150E	7.4000
IBM 3081 K	2.1000
IBM 4381-12	1.2000
IBM PC (w/ 8087)	0.0069

* Millions of Floating Point Operations Per Second

Table A2 (from RISC System/6000 and MIPS Staff) shows the performance of some recently introduced computers. The IBM RS/6000 is IBM's very aggressive entry into the workstation market. It's entry level price in the academic market is less than $10,000. The IBM PS/2 Model 70 with 486/25 Power Platform is one of the first computers with the Intel 486. The 486 is the most recent addition to the line of microprocessors that began with the Intel 8086 used in the IBM PC.

TABLE A2

COMPUTER	MFLOPS*
Vax 9000	18.0000
IBM RS/6000	7.3000
IBM PS/2 Model 70 w/ 486/25 Power Platform	0.9590

* Millions of Floating Point Operations

Finally, Table A3 (from Bell) shows some predictions for future computer performance.

TABLE A3

COMPUTER	MFLOPS*
Leading Microprocessor in 1992	20.0000
With Vector Processor	200.0000

* Millions of Floating Point Operations

One can see that floating point performance has risen recently and is posed to rise substantially in the near future. The new IBM 486 personal computer has about half the performance of their premier mainframe from the early 1980's (the 3081), while the RS/6000 approaches the first Cray in power. As a quick test of the RS/6000, a million evaluations of the Judge function were run on it. This required 35 seconds of execution time, while the X-MP took 21 seconds. Along with the turnaround time on mainframes and supercomputers, the RS/6000 is extremely competitive. Further, it is just a harbinger of future machines.

Appendix B: Simulated Annealing Algorithm

```
SET initial parameters and values
   set C          (controls how fast V adjusts)
   set X          (starting values for model parameters)
   set V          (should cover the entire range of interest in X)
   set ε          (convergence criteria)
   set r_T        (temperature reduction factor)
   set T_0        (initial temperature)
   set N_ε        (# times ε tolerance is achieved before end)
   set N_S        (# times through fcn. before V adj.)
   set N_T        (# times through N_S loop before T reduction)

CALCULATE f(X)
   X_opt = X
   f_opt = f

DO UNTIL convergence
    DO N_T times
        DO N_S times
            DO i = 1, n
                x_i' = x_i + r·v_i {r is uniform random # on [-1,1]}
                CALCULATE f'(X')
                IF f' ≤ f THEN apply Metropolis criteria & IF accepted X = X'
            & f = f'
                IF f' > f THEN X = X' & f = f'
                IF f' > f_opt THEN X = X', f = f', X_opt = X', & f_opt = f'
            END DO
        END DO
    ADJUST V such that half of all trials accepted
END DO

IF change in f_opt < ε last N_ε iterations & |f - f'| < ε
        THEN REPORT X_opt, f_opt & V
        STOP
ELSE
X = X_opt {start on current best optimum}
T = r_T·T {reduce T}

CONTINUE
```

ON EFFICIENT EXACT MAXIMUM LIKELIHOOD ESTIMATION OF HIGH-ORDER ARMA MODELS

STEFAN MITTNIK

Department of Economics, State University of New York at Stony Brook, Stony Brook, New York 11794-4384

1. Introduction

The evaluation of the exact likelihood function of an autoregressive moving average (ARMA) process via Kalman filtering requires the specification of the unconditional covariance matrix of the state vector. Gardner et al. (1980) suggest a derivation which is widely used in empirical work. It involves finding the solution to a Lyapunov-type equation and amounts to solving a system of linear equations. Depending on the order of the ARMA model, the number of unknowns in this system can be rather large, since it corresponds to the number of unique elements in the unconditional state covariance matrix. An alternative algorithm for constructing the covariance matrix for univariate models which considers the specific structure of the state vector is given in Jones (1980). Using a state space representation that is different from the one in Gardner et al. and commonly used in the systems literature,[1] Mélard (1984) and Shea (1987) apply the Chandrasekhar version of the Kalman filter (see Morf et al., 1974) to compute the exact likelihood of ARMA models. This approach is computationally more efficient and requires only the first block column of the state covariance matrix rather than the full matrix. The Chandrasekhar algorithm is, however, not applicable in exact maximum likelihood procedures designed for missing-data situations as suggested by Ansley and Kohn (1983), for example.

In contrast to Gardner et al., where the solution of the Lyapunov equation yields directly the desired quantities, Jones (1980), Mélard (1984), Shea (1987), and Mittnik (1990a) first derive the theoretical autocovariances from the underlying ARMA parameters and construct from those the unconditional state covariance matrix. In the multivariate case, the calculation of the theoretical autocovariance is usually accomplished with algorithms suggested in Nicholls and Hall (1979), Ansley (1980) or Kohn and Ansley (1982). But even with the improvements in Mittnik (1990b), the estimation of models with high-order autoregressive parts will be costly computationally.

[1] This state space representation, which has also been adopted by Pearlman (1980), will not be considered here. For a brief discussion of the two alternative representations and a nonrecursive approach for deriving the unconditional state covariance matrix of the "systems representation" see Mittnik (1990a).

Hans M. Amman et al. (eds), Computational Economics and Econometrics, 33-42.

Here we consider the more commonly adopted state space representation of Gardner et al. Deriving a closed-form expression for the full unconditional state covariance matrix, we develop a new, recursive procedure for computing the elements of the matrix. Moreover, we utilize the Levinson-type algorithm proposed in Mittnik (1990c) for computing the theoretical autocovariances of multiple ARMA processes. A combination of these two procedures is especially advantageous for high-order models.

2. Exact Maximum Likelihood Estimation via Kalman Filtering

Given N observations of a zero-mean, covariance stationary time series, y_i ($i=1,2,...,N$), we want to estimate the ARMA(p,q) model

$$\Phi(L)y_t = \Theta(L)\varepsilon_t, \tag{1}$$

where $E(\varepsilon_t)=0$ and $E(\varepsilon_s\varepsilon_t')=\delta_{st}\Sigma$, via exact maximum likelihood methods. The roots of the matrix polynomials $\Phi(L)=I-\Phi_1L-...-\Phi_pL^p$ and $\Theta(L)=\Theta_0+\Theta_1L+...+\Theta_qL^q$ are assumed to lie outside the unit circle, guaranteeing covarience stationarity and invertibility. Note that either Θ_0 or Σ could be set equal to the identity matrix. For the sake of a general presentation, we allow $\Theta_0\neq I$ and $\Sigma\neq I$.

A computationally efficient procedure for obtaining exact maximum likelihood estimates is provided by the Kalman filtering recursions. To apply the Kalman filter recursions, we cast the ARMA model (1) in the following state space form (c.f. Gardner et al., 1980)

$$z_t = Az_{t-1} + B\varepsilon_t \tag{2a}$$

$$y_t = Cz_t, \tag{2b}$$

with $C=[I\ 0...0]$ ($m\times mr$),

$$A = \begin{bmatrix} \Phi_1 & I & 0 & \cdots & 0 \\ \Phi_2 & 0 & I & \cdots & 0 \\ \vdots & & & \ddots & \vdots \\ \Phi_{r-1} & & & & I \\ \Phi_r & 0 & & \cdots & 0 \end{bmatrix} \quad \text{and} \quad B = \begin{bmatrix} \Theta_0 \\ \Theta_1 \\ \vdots \\ \Theta_{r-1} \\ \Theta_r \end{bmatrix}$$

where r=max(p,q+1) and Φ_i=0 (i=p+1,p+2,...,r), if r>p, or Θ_i=0 (i=q+1,q+2,...,r), if r>q.

Let Y denote the data vector and β the vector containing the parameters of model (1). Then, the log likelihood associated with Y and β is given by

$$\ln L(Y;\beta) = \text{const.} - \frac{1}{2}\sum_{t=1}^{N} \ln|\Lambda_t| - \frac{1}{2}\sum_{t=1}^{N} e_t'\Lambda_t^{-1}e_t. \tag{3}$$

The quantities e_t and Λ_t in (3) can be derived from the Kalman filtering recursions

$$e_t = y_t - Cz_{t|t-1} \tag{4a}$$

$$\Lambda_t = CP_{t|t-1}C' \tag{4b}$$

$$z_t = z_{t|t-1} + P_{t|t-1}\Lambda_t^{-1}e_t \tag{4c}$$

$$P_t = P_{t|t-1} - P_{t|t-1}C'\Lambda_t^{-1}CP_{t|t-1} \tag{4d}$$

$$z_{t|t-1} = Az_{t-1} \tag{4e}$$

$$P_{t|t-1} = AP_{t-1} + B\Sigma B'. \tag{4f}$$

The initial conditions are specified by setting the initial state vector, $z_{1|0}$, and the initial state covariance matrix, $P \triangle P_{1|0}$, to their respective unconditional expectations, i.e.,

$$z_{1|0} = \mathrm{E}(z_t) = 0 \tag{5}$$

and

$$P \triangle P_{1|0} = \mathrm{E}(z_t z_t'). \tag{6}$$

It follows from (2a) that

$$E(z_t z_t') = E[(Az_{t-1}+B\varepsilon_t)(Az_{t-1}+B\varepsilon_t)'].$$

Because $E(z_t z_t')=E(z_{t-1}z_{t-1}')$, $E(z_{t-1}\varepsilon_t')=0$ and $E(\varepsilon_t\varepsilon_t')=\Sigma$, we have

$$P = APA' + B\Sigma B'. \tag{7}$$

Relationship (7) is a referred to as a (discrete) Lyapunov equation. Gardner et al. compute P by vectorizing (7), yielding the system of $(mr)^2$ linear equations given by

$$(I - A\otimes A)vec(P) = vec(B\Sigma B'), \tag{8}$$

and solve for vec(P). Due to the symmetry of P, this approach requires us to solve for $mr(mr+1)/2$ unknowns. Since the log likelihood is maximized with respect to β using iterative, nonlinear optimization algorithms, (7) has to be solved at the beginning of each iteration. As a consequence, the computation of the initial state covariance can add substantially to the computational burden of exact maximum likelihood estimation procedures--especially, for high-order ARMA models.

3. Specification of the Unconditional State Covariance Matrix

Rather than solving Lyapunov equation (7), we will utilize the structure of the state vector, z_t, implied by state space representation (2). Let $z_{k,t}$ denote the k-th $m\times 1$ subvector of z_t, then

$$z_{k,t} = \sum_{i=k}^{p} \Phi_i y_{t+k-1-i} + \sum_{j=k-1}^{q} \Theta_j \varepsilon_{t+k-1-j}, \qquad k=1,2,\ldots,r. \tag{9}$$

Rewriting (9) in matrix form yields

$$z_t = \Phi y + \Theta u, \tag{10}$$

where $y=(y'_{t-1},y'_{t-2},...,y'_{t-r})'$, $u=(\varepsilon'_t,\varepsilon'_{t-1},...,\varepsilon'_{t-r+1})'$, and the $mr \times mr$ block-Hankel matrices Φ and Θ are given by

$$\Phi = \begin{bmatrix} \Phi_1 & \Phi_2 & \dots & \Phi_r \\ \Phi_2 & \Phi_3 & & 0 \\ \vdots & & & \vdots \\ \Phi_r & 0 & \dots & 0 \end{bmatrix} \quad \text{and} \quad \Theta = \begin{bmatrix} \Theta_0 & \Theta_1 & \dots & \Theta_{r-1} \\ \Theta_1 & \Theta_2 & & 0 \\ \vdots & & & \vdots \\ \Theta_{r-1} & 0 & \dots & 0 \end{bmatrix},$$

respectively.

Using (11), the unconditional state covariance matrix can be expressed as

$$P = \mathrm{E}[(\Phi y+\Theta u)(\Phi y+\Theta u)']. \tag{11}$$

Defining $\Omega_y \triangleq \mathrm{E}(yy')$, $\Omega_u \triangleq \mathrm{E}(uu')$ and $\Omega_{yu} \triangleq \mathrm{E}(yu')$, we obtain

$$P = \Phi\Omega_y\Phi' + \Theta\Omega_u\Theta' + \Phi\Omega_{yu}\Theta' + \Theta\Omega'_{yu}\Phi'. \tag{12}$$

Thus, P is simply the quadratic form of the ARMA parameters, collected in matrix $[\Phi,\Theta]$, weighted by the covariance matrix of the joint process $(y',u')'$. Denoting the unconditional covariance matrix of $(y',u')'$ by

$$\Omega \triangleq \mathrm{E}\left(\begin{bmatrix} y \\ u \end{bmatrix}(y' \; u')\right) = \begin{bmatrix} \Omega_y & \Omega_{yu} \\ \Omega'_{yu} & \Omega_u \end{bmatrix},$$

we have,

$$P = [\Phi \;\; \Theta]\Omega \begin{bmatrix} \Phi' \\ \Theta' \end{bmatrix}. \tag{13}$$

This is a result we expect, since the initial state vector is a linear combination of the presample realizations, $y_0,...,y_{-r+1},\varepsilon_0,...,\varepsilon_{-r+1}$, whose covariance matrix is given by Ω. Hence, the problem of specifying P is reduced to finding the covariance of the vector of presample realizations. The latter is a necessity in exact maximum likelihood procedures not based on the Kalman filtering as are, for example, described in Newbold (1974), Ansley (1979), and Nicholls and Hall (1979).

The submatrices of Ω are as follows. The variance of u is given by

$$\Omega_u = I_r \otimes \Sigma \triangleq \Sigma_r, \tag{14}$$

where I_n denotes an $n \times n$ identity matrix. The covariance matrix Ω_{yu} is specified by defining $\Psi_i = C_i \Sigma$ $(i=0,1,...,r-2)$, where C_i represents the i-th coefficient of the moving average representation of (1), i.e., $y_t = \Phi^{-1}(L)\Theta(L)\varepsilon_t = C(L)\varepsilon_t$, with $C(L) = C_0 + C_1 L + C_2 L^2 + ...$ Then,

$$\Omega_{yu} = \begin{bmatrix} 0 & \Psi_0 & \Psi_1 & \cdots & \Psi_{r-2} \\ 0 & & \Psi_0 & & \Psi_{r-3} \\ \vdots & & & \ddots & \vdots \\ & & & & \Psi_0 \\ 0 & & \cdots & & 0 \end{bmatrix}.$$

The coefficient of the moving average representation can be computed recursively by (c.f. Mittnik, 1987)

$$C_0 = \Theta_0,$$

$$C_i = \begin{cases} \Theta_i + \sum_{j=1}^{\min(i,p)} \Phi_j C_{i-j}, & \text{if } i \le q \\ \sum_{j=1}^{\min(i,p)} \Phi_j C_{i-j}, & \text{if } i > q \end{cases} \qquad i=1,2,.... \tag{15}$$

The remaining problem for computing P is the specification of submatrix Ω_y, which is given by the symmetric block-Toeplitz matrix

$$\Omega_y = \begin{bmatrix} \Gamma_0 & \Gamma_1' & \cdots & \Gamma_{r-1}' \\ \Gamma_1 & \Gamma_0 & & \Gamma_{r-2}' \\ \vdots & & \ddots & \vdots \\ \Gamma_{r-1} & \Gamma_{r-2} & \cdots & \Gamma_0 \end{bmatrix},$$

where $\Gamma_i \triangleq E(y_{t-i} y_t')$ $(i=0,\pm1,\pm2,...)$ represents the autocovariance sequence of process (1) satisfying $\Gamma_i = \Gamma_{-i}'$. Given the theoretical autocovariances Γ_i

(i=0,1,...,r-1) the unconditional state covariance matrix can be calculated from (11).

Several procedures for deriving the theoretical autocovariances for vector ARMA(p,q) processes have been proposed. Nicholls and Hall (1979) present a method for computing the autocovariances $\Gamma_0,...,\Gamma_p$ which requires us to solve a system of $m^2(p+1)$ linear equations. Making use of the symmetry of Γ_0, Ansley (1980) reduces the dimension to $m^2(p+1)-m(m-1)/2$. Kohn and Ansley (1982) achieve an improvement in computational efficiency by observing that Γ_p can be eliminated from the system of linear equations, reducing the problem to solving for $m^2p-m(m-1)/2$ unknowns. A more efficient algorithm is proposed in Mittnik (1990b). By deriving a closed-form expression for the initial autocovariances the latter approach provides insight into the construction of the coefficient matrix, allowing us to partition the coefficient matrix such that only (a triangular part of) Γ_0 and autocovariances Γ_1 through Γ_h, where $h=[(p-1)/2]$, if $p>1$, and $h=1$, if $p=1$, have to be derived simultaneously. Given $\Gamma_0,...,\Gamma_h$, the autocovariances $\Gamma_{h+1},...,\Gamma_{p-1}$ can be computed recursively.

Finally, an algorithm for calculating theoretical autocovariances that is especially advantageous for models with high-order autoregressive components and/or a large number of variables is proposed in Mittnik (1990c). Transforming the coefficient matrix of the equation system, such that it has a Hankel-plus-Toeplitz structure, enables us to use a computationally efficient block Levinson-type procedure for this task.

4. Recursive Computation of *P*

Rewriting the submatrices Ω_y, Ω_{yu} and Ω_u in terms of the actual ARMA orders p and q rather than their maximum, r, yields

$$\Omega_y = \begin{bmatrix} \Gamma_0 & \Gamma_1' & \cdots & \Gamma_{p-1}' \\ \Gamma_1 & \Gamma_0 & & \Gamma_{p-2}' \\ \vdots & & \ddots & \vdots \\ \Gamma_{p-1} & \Gamma_{p-2} & \cdots & \Gamma_0 \end{bmatrix}, \text{ if } p>q, \qquad \Omega_y = \left[\begin{array}{cccc|c} \Gamma_0 & \Gamma_1' & \cdots & \Gamma_{p-1}' & \\ \Gamma_1 & \Gamma_0 & & \Gamma_{p-2}' & 0' \\ \vdots & & \ddots & \vdots & \\ \Gamma_{p-1} & \Gamma_{p-2} & \cdots & \Gamma_0 & \\ \hline & & 0 & & 0 \end{array}\right], \text{ if } p\le q;$$

$$\Omega_{yu} = \begin{bmatrix} 0 & \Psi_0 & \Psi_1 & \cdots & \Psi_{q-1} & | & \\ 0 & & \Psi_0 & & \Psi_{q-2} & | & \\ \vdots & & & \ddots & \vdots & | & 0' \\ 0 & & \cdots & 0 & \Psi_0 & | & \\ - & - & - & - & - & | & - \\ & & 0 & & & | & 0 \end{bmatrix}, \text{ if } p>q, \qquad \Omega_{yu} = \begin{bmatrix} 0 & \Psi_0 & \cdots & & \Psi'_{q-1} \\ 0 & 0 & \ddots & & \Psi'_{q-2} \\ \vdots & & & & \vdots \\ 0 & & \cdots & 0 & \Psi_0 \end{bmatrix}, \text{ if } p \le q;$$

and

$$\Omega_u = \begin{bmatrix} I_{q+1} \otimes \Sigma & 0 \\ 0 & 0 \end{bmatrix} \text{ if } p>q, \qquad \Omega_u = I_{q+1} \otimes \Sigma \triangle \Sigma_{q+1}, \text{ if } p \le q.$$

Given $\Gamma_0,\dots,\Gamma_{p-1}$ and $\Psi_0,\dots,\Psi_{q-1}$, the (block) elements of the initial state covariance matrix P, denoted by $P_{k,n}$, can be computed. Since P is symmetric, only the entries $P_{k,n}$ for $k=1,\dots,r$ and $n=1,\dots,k$ need to be derived to specify P completely. Let P^y, P^{yu}, P^{uy}, and P^u denote the additive components of P that are associated with submatrices Ω_y, Ω_{yu}, Ω_{uy}, and Ω_u, respectively, i.e., $P=P^y+P^{yu}+P^{uy}+P^u$ and

$$P_{k,n} = P^y_{k,n} + P^{yu}_{k,n} + P^{uy}_{k,n} + P^n_{k,n}. \tag{16}$$

The individual terms are calculated as follows:

$$P^y_{k,n} = \sum_{i=n}^{p} \left(\sum_{j=k}^{p} \Phi_j \Gamma_{n-k-i+j} \right) \Phi'_i, \tag{17a}$$

$$P^{yu}_{k,n} = \sum_{i=k}^{p} \Phi_i \left(\sum_{j=n+i-k}^{q} \Psi_{k-n-i+j} \Theta'_j \right), \tag{17b}$$

$$P^{uy}_{k,n} = P^{yu'}_{n,k} \tag{17c}$$

$$P^u_{k,n} = \sum_{i=0}^{q-k+1} \Theta_{k-1+i} \Sigma \Theta'_{n-1+i}, \tag{17d}$$

for $k=1,...,r$ and $n=1,...,k$. Note that the computation of Ψ_i ($i=0,...,q-1$) involves only the first recursion in (15), since $i<q$.

Upon examining (17), we observe that the (block) elements of the additive components P^y, P^{yu} and P^u can be computed recursively along their diagonals. Given, for example, entry (k+1,n+1) of P^y, we can calculate entry (k,n) by adding the appropriate terms to entry (k+1,n+1). The following recursive relationships can be established:

$$P^y_{k,n} = \begin{cases} 0, & \text{if } k>p \text{ and } n=k,k-1,...,1 \\ P^y_{k+1,n+1} + \Phi_k \sum_{i=n+1}^{p} \Gamma_{n-i}\Phi_i' + \left(\sum_{j=k}^{p} \Phi_j \Gamma_{j-k}\right)\Phi_n', & \text{if } k=p,p-1,...,1 \text{ and} \\ & n=k,k-1,...,1; \end{cases} \tag{18a}$$

$$P^{yu}_{k,n} = \begin{cases} 0, & \text{if } k>g \text{ and } n=k,k-1,...,1 \\ P^{yu}_{k+1,n+1} + \Phi_k \sum_{i=n}^{q} \Psi_{j-n}\Theta_j', & \text{if } k=g,g-1,...,max(g-n+1,1) \text{ and} \\ & n=q,q-1,...,1, \end{cases} \tag{18b}$$

where g=min(p,q), and

$$P^u_{k,n} = \begin{cases} 0, & \text{if } k>q+1 \text{ and } n=k,k-1,...,1 \\ P^u_{k+1,n+1} + \Theta_{k-1}\Sigma\Theta_{n-1}' & \text{if } k=q+1,q,...,q-n+2 \text{ and } n=k,k-1,...,1. \end{cases} \tag{18c}$$

Combining (16), (17c) and (18) enables us to compute the elements of P recursively. When applying the Chandrasekhar version of the Kalman filter, only the first block column of P is required. But as it turns out, all elements of P not contained in the first block column are given as intermediate results when the first block column is computed. Thus, the computation of the full unconditional state covariance matrix, which is required in the presence of missing data, involves the same computational burden as the derivation of the first block column.

5. Conclusions

We have presented a closed-form expression for the elements of the unconditional state covariance matrix of multivariate ARMA models cast in state space form. The results provide insight into the structure of the matrix and enable us to simplify exact maximum likelihood estimation procedures that are based on Kalman filtering techniques. An efficient recursive algorithm for deriving the complete unconditional state covariance matrix has been suggested. It is particularly useful when estimating multivariate, high-order ARMA models or when facing missing-data problems.

References

Ansley, C. F., 1979,"An algorithm for the exact likelihood of mixed autoregressive-moving average process", *Biometrika* 66, 59-65.

Ansley, C. F., 1980, "Computation of the theoretical autocovariance function for a vector ARMA process", *Journal of Statistical Computation and Simulation* 12, 15-24.

Ansley, C. F. and R. Kohn, 1983, "Exact likelihood of vector autoregressive-moving average processes with missing or aggregated data", *Biometrika* 70, 275-278.

Gardner G., A. C. Harvey and G. D. A. Phillips, 1980,"An algorithm for exact maximum likelihood estimation of ARMA models by means of the Kalman filter", *Applied Statistics* 29, 311-322.

Jones, R. H., 1980,"Maximum likelihood fitting of ARMA models to time series with missing observations", *Technometrics* 22, 389-395.

Kohn, R. and C. F. Ansley, 1982,"A note on obtaining the theoretical autocovariances of an ARMA process", *Journal of Statistical Computation and Simulation* 15, 273-283.

Mélard, G., 1984,"A fast algorithm for the exact likelihood of autoregressive-moving average models", *Applied Statistics* 33, 104-114.

Mittnik, S., 1987,"Non-recursive methods for computing the coefficients of the autoregressive and the moving-average representation of mixed ARMA processes", *Economics Letters* 23, 279-284.

Mittnik, S. 1990a,"Derivation of the unconditional state covariance matrix for exact maximum likelihood estimation of ARMA models", forthcoming in *Journal of Economic Dynamics and Control.*

Mittnik, S., 1990b,"Computation of theoretical autocovariances of multivariate ARMA models", *Journal of the Royal Statistical Society* B 52, 151-155.

Mittnik, S., 1990c,"Computing theoretical autocovariances of multivariate ARMA models using a block-Levinson Method", unpublished manuscript.

Morf, M., G.S. Sidu and T. Kailath, 1974,"Some new algorithms for recursive estimation in constant, linear, discrete-time systems", *IEEE Transactions on Automatic Control*, AC-19, 315-323.

Newbold, P., 1974, "The exact likelihood function for a mixed autoregressive-moving average process", *Biometrika* 61, 423-426.

Nicholls, D. and A. D. Hall, 1979,"The exact likelihood function of multivariate autoregressive-moving average models", *Biometrika* 66, 259-264.

Pearlman, J. G., 1980,"An algorithm for the exact likelihood of a high-order autoregressive-moving average process", *Biometrika* 67, 232-233.

Shea, B. L., 1987,"Estimation of multivariate time series", *Journal of Time Series Analysis* 8, 95-109.

EFFICIENT COMPUTATION OF STOCHASTIC COEFFICIENTS MODELS

I-LOK CHANG *

Department of Mathematics and Statistics, The American University, Washington DC 20016

CHARLES HALLAHAN

Economic Research Service, United States Department of Agriculture, Washington, DC 20005

and

P. A. V. B. SWAMY**

Monetary & Financial Studies Section, Federal Reserve Board, Washington, DC 20551

Abstract. Paige (1979) has produced a reliable computational approach for solving the generalized linear least squares problem. This approach is used here to produce a numerically stable algorithm for computing stochastic coefficients models. This algorithm is well suited to these models because the matrices appearing in them may be singular and the algorithm is made to handle any rank deficiency.

1. Introduction

The feasibility of estimating a model when all of its coefficients follow a vector stochastic process with no known parameters was established by Swamy and Tinsley (1980). Such a model is desirable because, unlike the traditional fixed coefficients model, it does not conceive the observations on a dependent variable "as determined by a mathematical formula with added random disturbance, which in mechanistic terms involves the action of substituting in a formula, throwing dice and performing a computation" (see Dempster 1988, p. 151). Activities like dice throwing offer, in themselves, no explanation of how data are generated. Also, stochastic coefficients models clarify the gaps in information which must be filled from external sources if useful results are to be obtained. The present paper attempts development of a numerically stable and efficient algorithm for computing Swamy and Tinsley's estimator.

* Supported by a grant from the Board of Governors of the Federal Reserve System.
** We are grateful to Professor David Belsley for valuable comments on an earlier draft of this paper. Views expressed in this paper are those of the authors and do not necessarily reflect those of the Board of Governors or the staff of the Federal Reserve System.

H.M. Amman et al. (eds), Computational Economics and Econometrics, 43–53.

2. Estimation of Stochastic Coefficients Models

2.1. A Stochastic Coefficients Model

The basic stochastic coefficients model can be represented in terms of the following equations (a)-(d):

(a) The observation equation is of the form

$$y_t = (x_{1t}, x_{2t}, \ldots, x_{Kt}) \begin{pmatrix} \beta_{1t} \\ \beta_{2t} \\ \vdots \\ \beta_{Kt} \end{pmatrix}$$

$$= \mathbf{x}_t' \boldsymbol{\beta}_t \quad (t = 1,2,\ldots,T), \tag{1}$$

where y_t is an observable scalar random variable, $\mathbf{x}_t$ is a K-vector of observable explanatory variables with its first element, x_{1t}, identically equal to 1, $\boldsymbol{\beta}_t$ is a K-vector of unobservable coefficients and the coefficient on x_{1t} represents the sum of a random intercept and the disturbance term.

(b) $\boldsymbol{\beta}_t$ is assumed to satisfy the equation

$$\boldsymbol{\beta}_t = \Pi \mathbf{z}_t + L\epsilon_t, \tag{2}$$

where Π is a $K \times m_1$ matrix of fixed coefficients, $\mathbf{z}_t$ is an m_1-vector of observable variables, L is a $K \times m_2$ known fixed matrix, and ϵ_t is an m_2-vector of random variables.

(c) $\{\epsilon_t\}$ is assumed to satisfy the difference equation

$$\epsilon_t = \Phi \epsilon_{t-1} + \mathbf{a}_t, \tag{3}$$

where Φ is an $m_2 \times m_2$ fixed matrix whose eigenvalues are less than one in absolute value and $\{\mathbf{a}_t\}$ is a sequence of uncorrelated m_2-vector variables with

$$E\mathbf{a}_t = 0, \quad E\mathbf{a}_t\mathbf{a}_s' = \begin{cases} \sigma_a^2 \Delta_a & \text{if } t = s \\ 0 & \text{if } t \neq s. \end{cases}$$

(d) Exact linear restrictions on Π may be expressed as

$$R vec(\Pi) = \mathbf{r}, \tag{4}$$

where R is a $q \times Km_1$ known matrix of rank $q < Km_1$, the notation $vec(\Pi)$ is used to denote the column stack of the matrix Π, and $\mathbf{r}$ is a known q-vector. Exact restrictions on Φ and Δ_a may be expressed similarly. Various fixed coefficients models considered by economists fall out as special cases of equations (1)-(3) (see Swamy, Conway and LeBlanc 1988).

Combining equations (1) and (2) reveals that the stochastic coefficients representation can be viewed as a fixed coefficients model where the form of the error term is very general:

$$y_t = (\mathbf{z}_t' \otimes \mathbf{x}_t') vec(\Pi) + u_t, \tag{5}$$

where $\otimes$ denotes the Kronecker product and $u_t = \mathbf{x}_t' L \epsilon_t$.

Assumption (3) implies that $E\epsilon_t = 0, E\epsilon_t\epsilon_t' = \sigma_a^2\Gamma_0 = \Phi\sigma_a^2\Gamma_0\Phi' + \sigma_a^2\Delta_a$, $E\epsilon_t\epsilon_{t-s}' = \Phi^s\sigma_a^2\Gamma_0$ and $E\epsilon_t\epsilon_{t+s}' = \sigma_a^2\Gamma_0\Phi'^s$. From these results we can easily guess the form of the covariance matrix of $\epsilon = (\epsilon_1', \epsilon_2', \ldots, \epsilon_T')'$. We denote $E\epsilon\epsilon'$ by $\sigma_a^2\Sigma_\epsilon$. Let $\mathbf{y} = (y_1, y_2, \ldots, y_T)'$ and let $\mathbf{u} = (u_1, u_2, \ldots, u_T)'$. Then the covariance matrix of $\mathbf{y}$ or $\mathbf{u}$ is

$$\sigma_a^2\Sigma_y = D_x(I_T \otimes L)\sigma_a^2\Sigma_\epsilon(I_T \otimes L')D_x', \tag{6}$$

where $D_x = diag[\mathbf{x}_1', \mathbf{x}_2', \ldots, \mathbf{x}_T']$ is a $T \times TK$ block diagonal matrix and I_T is the $T \times T$ identity matrix. Since $\Delta_a, \Gamma_0, \Sigma_\epsilon$ and Σ_y are symmetric non-negative definite matrices, they have factorizations

$$\begin{aligned} \Delta_a &= W_aW_a', \Gamma_0 = W_0W_0', \Sigma_\epsilon = W_\epsilon W_\epsilon', \quad \text{and} \\ \Sigma_y &= D_x(I_T \otimes L)W_\epsilon W_\epsilon'(I_T \otimes L')D_x' \end{aligned} \tag{7}$$

such that W_a, W_0, and

$$W_\epsilon = \begin{pmatrix} W_0 & 0 & \cdots & 0 \\ \Phi W_0 & W_a & \cdots & 0 \\ \Phi^2 W_0 & \Phi W_a & \cdots & 0 \\ \vdots & \vdots & & \vdots \\ \Phi^{T-1}W_0 & \Phi^{T-2}W_a & \cdots & W_a \end{pmatrix}$$

are lower triangular matrices. The matrix $D_x(I_T \otimes L)W_\epsilon$ is not lower triangular but can be so reduced by using the numerically stable QR decomposition (see Rao 1973, p. 21). Thus, we can find a lower triangular matrix W_y such that $\Sigma_y = W_yW_y'$. These triangular matrices are not unique unless their diagonal elements are required to be non-negative which is always possible if the corresponding covariance matrices are positive definite (see Rao 1973, p. 22).

In model (5), the parameter matrices are Π, Φ, and $\sigma_a^2 W_a$, all or some of whose elements may be unknown. Swamy and Tinsley exploit the information contained in the data on $y_t, \mathbf{x}_t, \mathbf{z}_t, t = 1, 2, \ldots, T$, and $L, R, \mathbf{r}$, to estimate these unknown elements and also to predict $\boldsymbol{\beta}_t$. However, it should be noted that if the ratio of T to the number of unknown parameters is not sufficiently large, then this information is not adequate to obtain precise and unique estimates of the unknown parameters. Although the problem of imprecision may be solved by considering several alternative prior constraints that are consistent with the data, the nonuniqueness problem cannot be solved because the unique, suitable prior distribution for the unknown parameters cannot be determined. The Kalman filter also gives the predictions of $\boldsymbol{\beta}_t$ but requires the strong assumptions that the parameter matrices of model (5) are completely known, including a prior distribution of $\boldsymbol{\beta}_t$ at $t = 0$, and that a white noise component can be separated from β_{1t} so that it is uncorrelated with the remaining components of $\boldsymbol{\beta}_t$ (see Priestley 1981, p. 814). These assumptions are rarely, if ever, true and hence avoided by Swamy and Tinsley.

2.2. Iterative Algorithm to Estimate Parameters

The following flow diagram represents the iterative scheme we employ to estimate the unknown parameters of equations (1)-(3):

$$\hat{\Phi}, \hat{\Delta}_a, \hat{\sigma}_a^2 \longrightarrow \hat{\Gamma}_0 \longrightarrow \hat{\Sigma}_\epsilon \longrightarrow \hat{\Sigma}_y \longrightarrow \hat{\Pi} \longrightarrow \{\hat{\epsilon}_t\} \longrightarrow \{\hat{\beta}_t\}$$

Repeat until "convergence"

The steps involved in this iterative scheme are enumerated below:

1. Starting with the initial values $\Sigma_\epsilon^{(0)} = I$ and $\Sigma_y^{(0)} = D_x(I_T \otimes L)(I_T \otimes L')D_x'$ implied by the initial values $\Phi = \Phi^{(0)} = 0$ and $\Delta_a = \Delta_a^{(0)} = I$, we apply Paige's method (see Kourouklis and Paige 1981) to the following equation to obtain initial estimates of $vec(\Pi)$ and σ_a^2.

$$\mathbf{y}_e = X_{ze}vec(\Pi) + \mathbf{u}_e, \tag{8}$$

where $\mathbf{y}_e = (\mathbf{y}', \mathbf{r}')'$, $X_{ze} = (\mathbf{z}_1 \otimes \mathbf{x}_1, \mathbf{z}_2 \otimes \mathbf{x}_2, \ldots, \mathbf{z}_T \otimes \mathbf{x}_T, R')'$, and $\mathbf{u}_e = (\mathbf{u}', 0')'$ is obtained by adding q zero rows to the vector $\mathbf{u}$. We denote $E\mathbf{u}_e\mathbf{u}_e'$ by $\sigma_a^2\Sigma_{ye} = \sigma_a^2 W_{ye}W_{ye}'$, where W_{ye} is a lower triangular matrix. Note that the covariance matrix of $\mathbf{u}_e$ is singular even when the covariance matrix of $\mathbf{u}$ is nonsingular, and Paige's method works even in the singular covariance matrix case. Furthermore, Paige's approach gives numerically reliable results because it replaces solution methods based on explicit regular or generalized inverses by numerically stable methods of solution based on QR decompositions, as shown by Kourouklis and Paige (1981) and Paige (1979). Suppose that $X_z = (\mathbf{z}_1 \otimes \mathbf{x}_1, \mathbf{z}_2 \otimes \mathbf{x}_2, \ldots, \mathbf{z}_T \otimes \mathbf{x}_T)'$ and let $\mathcal{M}(X_z, \Sigma_y^{(0)})$ denote the column space of the partitioned matrix in parentheses. In the case when $\Sigma_y^{(0)}$ is singular, the consistency condition stated, for example, by Rao (1973, p. 297), namely

$$\mathbf{y} \in \mathcal{M}(X_z, \Sigma_y^{(0)}) \text{ with probability } 1, \tag{9}$$

is important because if this condition does not hold, then model (5) with $\Sigma_y = \Sigma_y^{(0)}$ is not true. Paige's method gives an easy computational check on this condition.

2. Writing that $\mathbf{u}_e = W_{ye}\mathbf{v}_e$ with $E\mathbf{v}_e\mathbf{v}_e' = \sigma_a^2 I$, Kourouklis and Paige (1981, p. 621) proved that if condition (9) is true, then Paige's algorithm gives solutions $vec(\hat{\Pi}^{(0)})$ and $\hat{\mathbf{v}}_e^{(0)}$ which satisfy the equation

$$\mathbf{y}_e = X_{ze}vec(\hat{\Pi}^{(0)}) + W_{ye}^{(0)}\hat{\mathbf{v}}_e^{(0)}, \tag{10}$$

where $W_{ye}^{(0)}$ and $\Sigma_{ye}^{(0)} = W_{ye}^{(0)}W_{ye}^{\prime(0)}$ are, respectively, the initial values of W_{ye} and Σ_{ye} implied by the initial value $\Sigma_y^{(0)}$, $\hat{\mathbf{v}}_e^{(0)}$ is unique, and $vec(\hat{\Pi}^{(0)})$ need not be unique but will be if we take the solution $vec(\hat{\Pi}^{(0)})$ with minimum Euclidean norm. Rao (1978, pp. 296-301) has shown that a representation of $vec(\hat{\Pi}^{(0)})$ with minimum Euclidean norm is $[X'_{ze}(\Sigma_{ye}^{(0)} + X_{ze}BX'_{ze})^{-}X_{ze}]_m^{-} \cdot X'_{ze}(\Sigma_{ye}^{(0)} + X_{ze}BX'_{ze})_\ell^{-}\mathbf{y}_e$, where B is any symmetric matrix such that $rank(X_{ze}, \Sigma_{ye}^{(0)}) = rank(\Sigma_{ye}^{(0)} + X_{ze}BX'_{ze})$, and A_m^- and A_ℓ^- denote minimum norm and least squares generalized inverses of A, respectively. The matrix B can be null when $\Sigma_{ye}^{(0)}$ is nonsingular or $\mathcal{M}(X_{ze}) \subset \mathcal{M}(\Sigma_{ye}^{(0)})$.[1] When some of the nonzero eigenvalues of $\Sigma_{ye}^{(0)}$ are much smaller than others, any direct method of evaluating Rao's formula may introduce large numerical errors, as shown by Kourouklis and Paige (1981, p. 620). By contrast, $vec(\hat{\Pi}^{(0)})$ with minimum Euclidean norm is automatically and efficiently computed by solving a lower triangular system given by Paige's algorithm without finding B and generalized inverses. The initial estimator of σ_a^2 is given by $(\hat{\sigma}_a^{(0)})^2 = \hat{\mathbf{v}}_e^{\prime(0)}\hat{\mathbf{v}}_e^{(0)}/[rank(X_{ze}, W_{ye}) - rank(X_{ze})]$. Paige's algorithm also gives an estimate of the covariance matrix of $vec(\hat{\Pi}^{(0)})$ and this estimate, corresponding to $vec(\hat{\Pi}^{(0)})$ with minimum Euclidean norm, is guaranteed to be a symmetric non-negative definite matrix.

3. The vector of residuals may be computed as $\hat{\mathbf{u}}_e^{(0)} = \mathbf{y}_e - X_{ze}vec(\hat{\Pi}^{(0)})$, the last q rows of which are zero. Let $J = (I_T, 0')'$, where 0 is the $q \times T$ matrix of zeroes. Then the first T rows of $\hat{\mathbf{u}}_e^{(0)}$ are given by

$$\hat{\mathbf{u}}^{(0)} = (I - X_z[X'_{ze}(\Sigma_{ye}^{(0)} + X_{ze}BX'_{ze})^{-}X_{ze}]^{-}X'_{ze}(\Sigma_{ye}^{(0)} + X_{ze}BX'_{ze})_\ell^{-}J)\mathbf{u}$$

$$= M_{11}^{(0)}D_x(I_T \otimes L)\epsilon, \tag{11}$$

where

$$M_{11}^{(0)} = (I - X_z[X'_{ze}(\Sigma_{ye}^{(0)} + X_{ze}BX'_{ze})^{-}X_{ze}]^{-}X'_{ze}(\Sigma_{ye}^{(0)} + X_{ze}BX'_{ze})_\ell^{-}J)$$

and $\mathbf{u} = D_x(I_T \otimes L)\epsilon$. Now applying the criterion of minimum average mean square error linear unbiased prediction (see Swamy et al. 1989, p. 6) to equation (11) gives

$$\hat{\epsilon}^{(0)} = \Sigma_\epsilon^{(0)}(I_T \otimes L')D'_xM_{11}^{\prime(0)}[M_{11}^{(0)}\Sigma_y^{(0)}M_{11}^{\prime(0)}]^{-}\hat{\mathbf{u}}^{(0)}. \tag{12}$$

Equation (12) reduces to Swamy and Tinsley's (1980, p. 116) equation (4.10) if $\Sigma_y^{(0)}$ is nonsingular and equation (4) is not present. Partition the predictors as $\hat{\epsilon}^{(0)} = (\hat{\epsilon}_1^{\prime(0)}, \hat{\epsilon}_2^{\prime(0)}, \ldots, \hat{\epsilon}_T^{\prime(0)})'$, where $\hat{\epsilon}_t^{(0)}$ is an m_2- vector. Premultiplying $\hat{\epsilon}_t^{(0)}$ by $\mathbf{x}'_tL$ gives

$$D_x(I_T \otimes L)\hat{\epsilon}^{(0)} = (\mathbf{x}'_1L\hat{\epsilon}_1^{(0)}, \mathbf{x}'_2L\hat{\epsilon}_2^{(0)}, \ldots, \mathbf{x}'_TL\hat{\epsilon}_T^{(0)})'. \tag{13}$$

[1] A proof of the consistency of $vec(\hat{\Pi}^{(0)})$ is given by Havenner and Swamy (1981).

4. Consider

$$\begin{pmatrix} w_2^{(0)}\mathbf{x}_2'L\hat{\epsilon}_2^{(0)} \\ w_3^{(0)}\mathbf{x}_3'L\hat{\epsilon}_3^{(0)} \\ \vdots \\ w_T^{(0)}\mathbf{x}_T'L\hat{\epsilon}_T^{(0)} \end{pmatrix} = \begin{pmatrix} w_2^{(0)}(\hat{\epsilon}_1'^{(0)} \otimes \mathbf{x}_2'L) \\ w_3^{(0)}(\hat{\epsilon}_2'^{(0)} \otimes \mathbf{x}_3'L) \\ \vdots \\ w_T^{(0)}(\hat{\epsilon}_{T-1}'^{(0)} \otimes \mathbf{x}_T'L) \end{pmatrix} vec(\hat{\Phi}^{(1)}) + \begin{pmatrix} \mathbf{x}_2'L\hat{\mathbf{a}}_2^{(0)}w_2^{(0)} \\ \mathbf{x}_3'L\hat{\mathbf{a}}_3^{(0)}w_3^{(0)} \\ \vdots \\ \mathbf{x}_T'L\hat{\mathbf{a}}_T^{(0)}w_T^{(0)} \end{pmatrix}, \tag{14}$$

where $w_t^{(0)} = (\mathbf{x}_t'L\Delta_a^{(0)}L'\mathbf{x}_t)^{-1/2}$. The restriction that Φ is diagonal may be imposed by replacing the Kronecker products in equation (14) by their Hadamard products. When Φ is nondiagonal, we apply the Levenberg-Marquardt method to equation (14) to obtain an initial estimate of Φ subject to the restriction that the eigenvalues of Φ are less than one in absolute value. Details of this procedure are relegated to the Appendix. When Φ is diagonal, we apply Lawson and Hanson's (1974) algorithm based on the Kuhn-Tucker method to equation (14) to estimate Φ subject to the restriction that the diagonal elements of Φ are less than 1 in absolute value.

5. Consider

$$(1/(\hat{\sigma}_a^{(0)})^2)\begin{pmatrix} \mathbf{x}_2'L(\hat{\epsilon}_2^{(0)} - \hat{\Phi}^{(1)}\hat{\epsilon}_1^{(0)})(\hat{\epsilon}_2^{(0)} - \hat{\Phi}^{(1)}\hat{\epsilon}_1^{(0)})'L'\mathbf{x}_2 \\ \mathbf{x}_3'L(\hat{\epsilon}_3^{(0)} - \hat{\Phi}^{(1)}\hat{\epsilon}_2^{(0)})(\hat{\epsilon}_3^{(0)} - \hat{\Phi}^{(1)}\hat{\epsilon}_2^{(0)})'L'\mathbf{x}_3 \\ \vdots \\ \mathbf{x}_T'L(\hat{\epsilon}_T^{(0)} - \hat{\Phi}^{(1)}\hat{\epsilon}_{T-1}^{(0)})(\hat{\epsilon}_T^{(0)} - \hat{\Phi}^{(1)}\hat{\epsilon}_{T-1}^{(0)})'L'\mathbf{x}_T \end{pmatrix} = \begin{pmatrix} \mathbf{x}_2'L \otimes \mathbf{x}_2'LW_a^{(0)} \\ \mathbf{x}_3'L \otimes \mathbf{x}_3'LW_a^{(0)} \\ \vdots \\ \mathbf{x}_T'L \otimes \mathbf{x}_T'LW_a^{(0)} \end{pmatrix} vec(\hat{W}_a'^{(1)}) + \text{ an error vector}, \tag{15}$$

where $W_a^{(0)}$ is the initial value of W_a implied by $\Delta_a^{(0)}$. The diagonality restriction on Δ_a may be imposed by replacing the Kronecker products in equation (15) by their Hadamard products. We apply Lawson and Hanson's (1974) algorithm based on the Kuhn-Tucker method to equation (15) to obtain an initial estimate of W_a subject to the restriction that the diagonal elements of W_a are non-negative. Note that the estimate $\hat{\Delta}_a^{(1)} = \hat{W}_a^{(1)}\hat{W}_a'^{(1)}$ of Δ_a is guaranteed to be non-negative definite.

6. Knowing $\hat{\Phi}^{(1)}$ and $\hat{\Delta}_a^{(1)}$, we can determine $\hat{\Gamma}_0^{(1)}$ from the following equation:

$$\hat{\Gamma}_0^{(1)} = \hat{\Phi}^{(1)}\hat{\Gamma}_0^{(1)}\hat{\Phi}'^{(1)} + \hat{\Delta}_a^{(1)}, \tag{16}$$

where $\hat{\Phi}^{(1)}$ is known to have eigenvalues that are less than 1 in absolute value and $\hat{\Delta}_a^{(1)}$ is known to be non-negative definite. The exact solution to equation (16) is expressible as the convergent infinite series

$$\hat{\Gamma}_0^{(1)} = \hat{\Delta}_a^{(1)} + \hat{\Phi}^{(1)}\hat{\Delta}_a^{(1)}\hat{\Phi}'^{(1)} + (\hat{\Phi}^{(1)})^2\hat{\Delta}_a^{(1)}(\hat{\Phi}'^{(1)})^2 + \cdots \tag{17}$$

From this equation we can see that the non-negative definiteness of $\hat{\Delta}_a^{(1)}$ guarantees the non-negative definiteness of the exact solution of equation (16). However, when the maximum modulus of the eigenvalues of $\hat{\Phi}^{(1)}$ is near 1, the rate of convergence of the series (17) is so slow as to make the evaluation of (17) impractical. Therefore, we follow an alternative method based on the equation

$$(I - \hat{\Phi}^{(1)} \otimes \hat{\Phi}^{(1)})vec(\hat{\Gamma}_0^{(1)}) = vec(\hat{\Delta}_a^{(1)}), \tag{18}$$

where the identity matrix is of order $m_2^2 \times m_2^2$. To ensure the symmetry of $\hat{\Gamma}_0^{(1)}$, accomplishing at the same time a reduction in the number of operations, the columns of $(I - \hat{\Phi}^{(1)} \otimes \hat{\Phi}^{(1)})$ corresponding to the ijth and jith off-diagonal elements of $\hat{\Gamma}_0^{(1)}$ are added together. The resulting system which has m_2^2 linear equations in $m_2(m_2+1)/2$ unknowns, can be solved by using the QR linear least squares procedure. Let $\hat{\Gamma}_0^{(1)}(0)$ denote the initial solution of equation (18) given by this procedure. Although $\hat{\Gamma}_0^{(1)}(0)$ may be close to $\hat{\Gamma}_0^{(1)}$ in (17), it may not be non-negative definite. We use the following fixed-point iteration until convergence to improve the accuracy of $\hat{\Gamma}_0^{(1)}(0)$:

$$\hat{\Gamma}_0^{(1)}(i+1) = \hat{\Phi}^{(1)}\hat{\Gamma}_0^{(1)}(i)\hat{\Phi}'^{(1)} + \hat{\Delta}_a^{(1)}, \quad i = 0, 1, 2, \dots\ , \tag{19}$$

where $\hat{\Gamma}_0^{(1)}(i+1)$ is obtained by using $\hat{\Gamma}_0^{(1)}(i)$ in place of $\hat{\Gamma}_0^{(1)}$ used on the right-hand side of equation (16). Write

$$\hat{\Gamma}_0^{(1)}(0) = \hat{\Gamma}_0^{(1)} + Error(0), \tag{20}$$

where $\hat{\Gamma}_0^{(1)}$ is the exact solution in (17) and $Error(0)$ is the error in $\hat{\Gamma}_0^{(1)}(0)$. Substituting (20) into (19) for $i = 0$ gives

$$\hat{\Gamma}_0^{(1)}(1) = \hat{\Gamma}_0^{(1)} + \hat{\Phi}^{(1)}Error(0)\hat{\Phi}'^{(1)} \ \ [\text{by}(16)]. \tag{21}$$

If we substitute (21) into (19) for $i = 1$, then

$$\hat{\Gamma}_0^{(1)}(2) = \hat{\Gamma}_0^{(1)} + (\hat{\Phi}^{(1)})^2 Error(0)(\hat{\Phi}'^{(1)})^2 \ \ [\text{by}(16)]. \tag{22}$$

If we repeat this substitution n times, then we get

$$\hat{\Gamma}_0^{(1)}(n) = \hat{\Gamma}_0^{(1)} + (\hat{\Phi}^{(1)})^n Error(0)(\hat{\Phi}'^{(1)})^n. \tag{23}$$

The second term on the right-hand side of equation (23) is virtually a null matrix for sufficiently large n because the eigenvalues of $\hat{\Phi}^{(1)}$ have moduli less than 1. The iteration implied by equation (19) is terminated when the following condition is satisfied:

$$||\hat{\Gamma}_0^{(1)}(i+1) - \hat{\Gamma}_0^{(1)}(i)|| \leq \eta||\hat{\Gamma}_0^{(1)}(i+1)||,$$

where η is the magnitude of a machine zero and $|| \cdot ||$ denotes the supremum norm of a matrix. Now an estimate $\hat{W}_0^{(1)}$ of W_0 is more easily derived from the Cholesky factorization of $\hat{\Gamma}_0^{(1)}$.

7. Substituting $\hat{\Phi}^{(1)}, \hat{W}_a^{(1)}$ and $\hat{W}_0^{(1)}$ into W_ϵ gives $\hat{W}_\epsilon^{(1)}$ and substituting $\hat{W}_\epsilon^{(1)}$ into Σ_y shown in (7) gives $\hat{\Sigma}_y^{(1)}$, which is possibly singular or ill-conditioned even when Σ_y is nonsingular. From $D_x(I_T \otimes L)\hat{W}_\epsilon^{(1)}$, the estimates $\hat{W}_y^{(1)}$ and $\hat{W}_{ye}^{(1)}$ of W_y and W_{ye}, respectively, may be easily calculated.

8. The procedure is then iterated by using (10) with $W_{ye}^{(0)}$ replaced by $\hat{W}_{ye}^{(1)}$ to obtain a revised estimate of $vec(\Pi)$. This may in turn be used to obtain revised predictions of ϵ from (12) and revised estimates of Φ, W_a, W_0, and W_y from (14), (15), (23), and (7), respectively. The iteration may be continued until "convergence."

It would be interesting to know the circumstances in which this iterative algorithm is guaranteed to converge to a consistent and asymptotically efficient solution of the likelihood equations whenever such a solution exists.[2] The fact that an estimate of Σ_y can be singular in any iteration indicates that this is a difficult question. The logarithm of the likelihood function for model (5) need not be concave, and hence multiple solutions of the likelihood equations can exist. Also, the likelihood equations may not have a solution inside the boundary of the parameter space. More importantly, any direct method of solving the likelihood equations may introduce large numerical errors. In order to avoid these errors, Paige (1979) gives a numerically stable and efficient algorithm for solving the constrained least squares problem: minimize$_{vec(\Pi)}$ $\mathbf{v}_e'\mathbf{v}_e$ subject to $\mathbf{y}_e = X_{ze}vec(\Pi) + W_{ye}\mathbf{v}_e$. This problem involves only stable reductions to triangular systems of equations, which can be efficiently solved on a computer, and no restrictions on the ranks of X_{ze} and W_{ye} other than condition (9). It is equivalent to the generalized least squares problem if Σ_y is nonsingular and X_z has full column rank. Our problem is more complicated than Paige's because Σ_y is unknown. Since $\hat{\Sigma}_y^{(p)}$ obtained in the pth

[2] After discussing carefully existence and consistency of maximum likelihood estimators, Lehmann (1983, p. 435) gives a general method for constructing asymptotically efficient likelihood estimators.

iteration of our procedure is possibly singular, any formal proof of the convergence of our iterative algorithm must be based on conditions which imply that $\hat{\mathbf{v}}_e'^{(p)}\hat{\mathbf{v}}_e^{(p)}$ subject to the restriction $\mathbf{y}_e = X_{ze}vec(\hat{\Pi}^{(p)}) + \hat{W}_{ye}^{(p)}\hat{\mathbf{v}}_e^{(p)}$ is non-increasing on each iteration of Paige's algorithm and is strictly decreasing on any iteration such that $\hat{\mathbf{v}}_e'^{(p+1)}\hat{\mathbf{v}}_e^{(p+1)} < \hat{\mathbf{v}}_e'^{(p)}\hat{\mathbf{v}}_e^{(p)}$. These conditions should also imply that a consistent and asymptotically efficient solution of the likelihood equations is a fixed point of the iterative Paige algorithm. Condition (9) should also be satisfied in each iteration of Paige's algorithm. If the algorithm does not converge, then parameter estimates which produce the most satisfactory forecasts of y in an out-of-sample period may be selected by screening several iterations of the algorithm. Also, in such cases, intelligent programs for searching alternative minimal restrictions that lead to convergence can be designed. It would be interesting to compare these restrictions with the arbitrary restrictions that reduce model (5) to fixed coefficients models of the conventional type. The latter restrictions give unique results even when unique results are not possible because they are designed to do so.

These difficulties may not arise in cases where asymptotic theory applies. In such cases, either the estimates $vec(\hat{\Pi}^{(2)})$, $\hat{\Phi}^{(2)}$, $\hat{W}_a^{(2)}$, and $\hat{\sigma}_a^{(2)}$ obtained in the 2nd iteration of the algorithm or Lehmann's (1983, Theorem 4.2, p. 435) one-step estimator based on the initial estimates $vec(\hat{\Pi}^{(2)})$, $\hat{\Phi}^{(2)}$, and $\hat{W}_a^{(2)}$ may be used because these estimators can be asymptotically efficient.

After choosing the $\hat{\epsilon}_t$, $\hat{\Phi}$, and $\hat{\Pi}$, the $\boldsymbol{\beta}_t$ may be predicted by $\hat{\boldsymbol{\beta}}_t = \hat{\Pi}\mathbf{z}_t + L\hat{\epsilon}_t$, and the out-of-sample values of y may be predicted by $\hat{y}_{T+s} = \mathbf{x}_{T+s}'\hat{\Pi}\mathbf{z}_{T+s} + \mathbf{x}_{T+s}'L\hat{\Phi}^s\hat{\epsilon}_T$, $s = 1, 2, \ldots$.

Appendix

Write equation (14) as $\mathbf{c} \approx Hvec(\hat{\Phi}^{(1)})$. We shall use two matrix norms: the Frobenius norm defined by $\|\Phi\|_F^2 = \sum_{i,j}\phi_{ij}^2$ and the usual Euclidean norm defined by $\|vec(\Phi)\|_E^2 = (vec(\Phi))'(vec(\Phi))$. In this case, $\|\Phi\|_F = \|vec(\Phi)\|_E$.

We seek the $\hat{\Phi}^{(1)}$ that minimizes the norm $\|\mathbf{c} - Hvec(\hat{\Phi}^{(1)})\|_E$ subject to the restriction that the eigenvalues of $\hat{\Phi}^{(1)}$ are less than 1 in absolute value. Since the computational time needed to find such a $\hat{\Phi}^{(1)}$ by use of nonlinear numerical methods is prohibitive, we use an algorithm that can give us a satisfactory approximate solution to the problem cheaply. We minimize

$$\|\mathbf{c} - Hvec(\hat{\Phi}^{(1)})\|_E^2 + \mu^2\|vec(\hat{\Phi}^{(1)})\|_E^2, \tag{24}$$

where μ is a non-negative real parameter, with respect to $vec(\hat{\Phi}^{(1)})$ and μ. According to the results of Morrison (1960) and Marquardt (1963),

1. if $vec(\hat{\Phi}^{(1)})(\mu)$ minimizes (24), then it also minimizes $\|\mathbf{c} - Hvec(\hat{\Phi}^{(1)})\|_E^2$ subject to the constraint that $\|vec(\hat{\Phi}^{(1)})\|_E \leq \|vec(\hat{\Phi}^{(1)})(\mu)\|_E$ and
2. $lim_{\mu\to\infty}\|vec(\hat{\Phi}^{(1)})(\mu)\|_E = 0$.

Minimization of (24) can be easily done by using the singular-value decomposition of H. Since H is a $(T-1)\times m_2^2$ matrix of rank, say r, we can find two orthogonal matrices P and Q so that

$$H = P\begin{pmatrix} S & 0 \\ 0 & 0 \end{pmatrix} Q' \\ = PS_eQ', \tag{25}$$

where P is of order $(T-1)\times(T-1)$, S is a $r \times r$ diagonal matrix with positive diagonal elements, Q is of order $m_2^2 \times m_2^2$ and $S_e = \begin{pmatrix} S & 0 \\ 0 & 0 \end{pmatrix}$ is of order $(T-1)\times m_2^2$. The diagonal elements of S are called the singular values of H. If $vec(\hat{\Phi}^{(1)})(\mu)$ minimizes (24), then it satisfies the equation

$$(H'H + \mu^2 I)vec(\hat{\Phi}^{(1)})(\mu) = H'\mathbf{c}. \tag{26}$$

Using the singular-value decomposition (25), we can write the solution of equation (26) as

$$vec(\hat{\Phi}^{(1)})(\mu) = Q(S_e'S_e + \mu^2 I)^{-1}S_e'P'\mathbf{c}. \tag{27}$$

Equation (27) proves Morrison and Marquardt's result 2 but not their result 1.

If $s_{\min}$ and $s_{\max}$ denote the smallest and the largest singular values of $\hat{\Phi}^{(1)}(\mu)$, respectively, and if λ_i is the ith eigenvalue of $\hat{\Phi}^{(1)}(\mu)$, then it can be shown that $s_{\min} \le \min_i |\lambda_i| \le \max_i |\lambda_i| \le s_{\max}$ and $\|\hat{\Phi}^{(1)}(\mu)\|_F \ge s_{\max} \ge \max_i |\lambda_i|$.

Note that the regular inverse on the right-hand side of equation (27) will not exist if $\mu = 0$ and $m_2^2 > (T-1)$. In such cases, we use a generalized inverse in place of the regular inverse. If $\hat{\Phi}^{(1)}(0)$ is defined and its maximum eigenvalue has modulus less than 1, then it is a permissible solution to equation (14). Otherwise, using a root-finding algorithm (a modified regula falsi routine), we find a root μ^* such that the maximum modulus of the eigenvalues of $\hat{\Phi}^{(1)}(\mu^*)$ is equal to 0.995 and take $\hat{\Phi}^{(1)}(\mu^*)$ as the solution to equation (14).

The solution $\hat{\Phi}^{(1)}(\mu^*)$ is not exactly equal to the optimal solution to (14) because the feasible set of (14) may have elements outside the hypersphere $\|vec(\hat{\Phi}^{(1)})\|_E \le \|vec(\hat{\Phi}^{(1)})(\mu^*)\|_E$. The advantages of the approximate solution are that it is easy to compute, maximum modulus of its eigenvalues is less than 1, and it solves (14) in the sense of Morrison and Marquardt's result 1. An additional advantage is that the elements of $\hat{\Phi}^{(1)}(\mu^*)$ are not likely to become unmanageably large because of the constraint in (24).

References

Dempster, A. P.: 1988, 'Employment discrimination and statistical science (with discussion)', *Statistical Science* **3**, 149–195.

Havenner, H. and P. A. V. B. Swamy: 1981, 'A random coefficient approach to seasonal adjustment of economic time series', *Journal of Econometrics* **15**, 177–209.

Kourouklis, S. and C. C. Paige: 1981, 'A constrained least squares approach to the general Gauss-Markov linear model', *Journal of the American Statistical Association* **76**, 620–625.

Lawson, C. L. and R. J. Hanson: 1974, *Solving Least Squares Problems*, Prentice Hall, Englewood Cliffs, New Jersey.

Lehmann, E. L.: 1983, *Theory of Point Estimation*, Wiley, New York.

Marquardt, D. W.: 1963, 'An algorithm for least-squares estimation of nonlinear parameters', *Journal of the Society for Industrial and Applied Mathematics* **11**, 431–441.

Morrison, D. D.: 1960, 'Methods for nonlinear least squares problems and convergence proofs, Tracking Programs and Orbit Determination', *Proceedings of Jet Propulsion Laboratory Seminar*, Pasadena, California, Cochairmen: Jack Lorell and F. Yagi, pp. 1–9.

Paige, C. C.: 1979, 'Computer solution and perturbation analysis of generalized linear least squares problems', *Mathematics of Computation* **33**, 171–183.

Priestley, M. B.: 1981, *Spectral Analysis and Time Series*, Academic Press, New York.

Rao, C. R.: 1973, *Linear Statistical Inference and its Applications*, 2nd ed., Wiley, New York.

Rao, C. R.: 1978, 'Choice of best linear estimators in the Gauss-Markoff model with a singular dispersion matrix', *Communications in Statistics, Part A – Theory and Methods*, **A7**, 1199–1208.

Swamy, P. A. V. B. and P. A. Tinsley: 1980, 'Linear prediction and estimation methods for regression models with stationary stochastic coefficients', *Journal of Econometrics* **12**, 103–142.

Swamy, P. A. V. B., R. K. Conway, and M. R. LeBlanc: 1988, 'The stochastic coefficients approach to econometric modeling, Part II: Description and motivation', *The Journal of Agricultural Economics Research* **40**, 21–30.

Swamy, P. A. V. B., R. K. Conway, and M. R. LeBlanc: 1989, 'The stochastic coefficients approach to econometric modeling, Part III: Estimation, stability testing, and prediction', *The Journal of Agricultural Economics Research* **41**, 4–20.

THE DEGREE OF EFFECTIVE IDENTIFICATION AND A DIAGNOSTIC MEASURE FOR ASSESSING IT

DAVID A. BELSLEY[1]

Abstract. In this paper, after some introductory motivation, we define the *degree of effective identification* for an equation (or subset of equations) from a simultaneous system of equations and then suggest a diagnostic measure for assessing its strength based on the measure of adequacy for signal-to-noise given in Belsley (1982).

Interesting computationally-related elements arise in this context. Because of the empirical basis to the measure for signal-to-noise, appropriate Monte Carlo experiments will be required to calibrate it in this use, and the design of these experiments is currently underway. Furthermore, it is of interest to realize that one is unlikely even to attempt to define a measure of this sort for the degree of effective identification without knowing in the back of one's mind that the computational power exists to provide the empirical input needed to bring it to life.

Motivation

Quite generally, an equation is unidentifiable, and therefore inestimable, when it is not possible to determine its parameters uniquely from knowledge of the conditional distribution of y given $X \in \Theta$, where Θ is the sample space of relevant outcomes on X.[2] But, even though an equation may be identifiable, there can be specific X, usually contained in a subset of Θ of measure zero, on which estimation is impossible. For example, in the event that there is exact collinearity, so that some variates in X are indistinguishable from linear combinations of the others, the otherwise identifiable least-squares estimator is not uniquely determined at that X, and the equation is inestimable. And further, in the event that there is near collinearity, so that X is of full rank but poorly-conditioned, the equation is estimable, but weakly so. Thus, there is a relation between the strength of the data and the degree of estimability, or, what we might call the "degree of effective identification," a relation that is exploited and pursued here.

Of course, an equation from a linear system of equations is either identifiable or not, due to zero restrictions, depending on whether the rank condition holds. But the rank condition reflects a property of the actual conditional distribution of the endogenous variates given the exogenous variates, and as such, it is equivalent to having an infinite sample. As such, it cannot inform us of any practical problems that one might encounter in estimating a given equation with a specific, finite data set. This point can be seen readily from the following simple example. Suppose the equation

[1] Boston College and MIT. This paper was funded in part through a guest scholarship at the Rheinisch-Westfälisches Institut für Wirtschaftsforschung, Essen. West Germany, and in part through NSF Grant SES–8420614.

[2] See Malinvaud (1970).

H.M. Amman et al. (eds), Computational Economics and Econometrics, 55–62.

$$y_1 = \gamma_1 + \gamma_2 y_2 + \beta_1 x_1 + \varepsilon \,, \qquad (1)$$

is part of a simultaneous system of equations, where the y_is are endogenous and x_1 is exogenous. Further, assume this equation is identifiable. Now suppose that x_2 is an exogenous variate excluded from this equation and that it is the only other exogenous variate in the system. All of the identifying information, then, is embodied in x_2. Now, if x_1 and x_2 just happen to be highly collinear in the particular data set available for estimating (1), it is clear that, although x_2 is excluded in theory, most of its information is already contained in x_1 — and indeed, in the limiting case of perfect correlation between x_1 and x_2, all of its information is so contained —, and so it is not very "excluded" in fact. Thus, relative to these data, the given equation is poorly identified in practice, and, the greater the degree of collinearity between x_1 and x_2, the less the degree of effective identification.

This could occur, for example, in a supply and demand analysis in which (1) represents the demand equation with endogenous variates price and quantity and an exogenous variate x_1 representing tastes. We suppose the accompanying supply equation contains the same two endogenous variates along with an exogenous variate x_2 representing weather. If, now, for whatever reasons, the particular data set were characterized by a high degree of collinearity between weather and tastes — perhaps consumer's optimism is affected by the amount of rain —, then these two effects cannot be well separated, and there is no real identifying information to allow estimation to proceed with precision.[3]

Correlation between x_1 and x_2 is not the only reason why there may be ineffective identifying information contained in the excluded exogenous variates. It may be the case, for example, that the noise inherent in the error term of the given equation is so great that the signal available from the identifying excluded exogenous variates is overshadowed. Again considering the familiar textbook case of supply and demand, one recalls that the demand equation becomes identifiable when its excluded exogenous variates are able to shift the supply equation strongly in relation to it, as in Exhibit 1a.

The dotted lines here indicate the bounds of probable variation, and hence values for p and q will tend to lie in the intersection of these two dotted regions. In the situation of Exhibit 1a, the data on p and q will be clustered in a narrow band around the demand equation, allowing it to be well estimated. If, however, the additive noise in the demand equation D is great, then even the presence of additional exogenous variation in the supply equation S need not permit effective identification of D for small samples, as seen in Exhibit 1b. Here, the data on p and q will be dispersed about the large area bounded by the dotted lines, displaying little of the shape of the demand curve, even though it is technically identifiable.

The degree of effective identification, then, has to do with whether or not the

[3] Sims (1980) provides a closely related concept when he writes, "When the strictly exogenous variables have low explanatory power, estimates of the endogenous-on-exogenous regressions are likely to be subject to great sampling error, and the identification may be said to be weak."

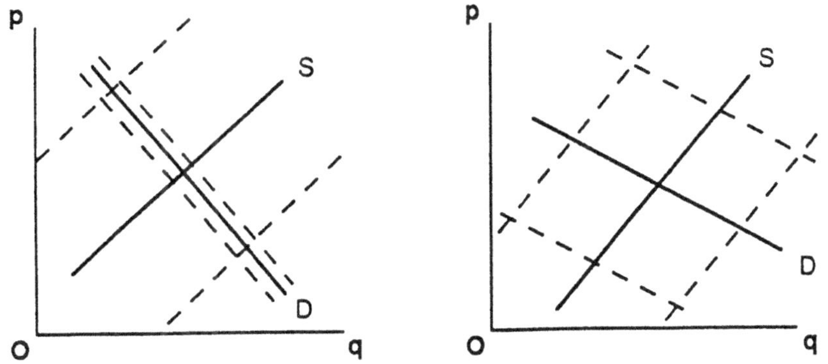

Fig. 1. **Exhibit 1** (a) strong effective identification of the demand function D, (b) weak effective identification of the demand function D.

excluded (identifying) exogenous variates provide adequate signal-to-noise, either by themselves or over and above that already embodied in the included exogenous variates, to allow estimation to proceed with precision, that is, to allow identification to take place as a practical matter. This suggests the relevance of the measure for signal-to-noise given in Belsley (1982) as a means for assessing the degree of effective identification.

The Diagnostic Measure

We begin with a very brief review of the relevant econometric background. Readers unfamiliar with the standard results regarding the identification and estimation of simultaneous-systems are referred to such texts as Theil (1971) or Johnston (1984). Readers unfamiliar with the test for the adequacy of signal-to-noise are referred to Belsley (1982).

Consider, then, the simultaneous system of linear equations

$$Y\Gamma + XB + U = 0\,, \tag{2}$$

where Y is an $n \times g$ matrix of g endogenous variates, X is an $n \times k$ matrix of k predetermined variates, U is an $n \times g$ matrix of temporally independent error terms with mean zero and nonsingular contemporaneous variance-covariance matrix Σ, and Γ and B are, respectively, $g \times g$ and $g \times k$ matrices of parameters to be estimated.

We shall be concerned with one of these equations, which we assume to be identifiable. When normalized and written with all zero restrictions, this equation is

$$y = Y_1 \gamma_1 + X_1 \beta_1 + u \,, \tag{3}$$

where y is the endogenous variate with respect to which the equation is normalized, Y_1 is the set of $g_1 - 1$ included endogenous variates, X_1 is the set of k_1 included exogenous variates, u is the appropriate column of U and, γ_1 and β_1 are the appropriate nonzero elements from Γ and B after normalization. In addition, there are the matrices of g_2 excluded endogenous variates Y_2 and k_2 excluded exogenous variates X_2, such that $X \equiv [X_1 \; X_2]$ and $Y \equiv [y \; Y_1 \; Y_2]$, $g = g_1 + g_2$, and $k = k_1 + k_2$.

Postmultiplying (2) by Γ^{-1}, which is assumed to exist, produces the reduced form

$$Y = X\Pi + V \,, \tag{4}$$

where $\Pi \equiv -B\Gamma^{-1}$ and $V \equiv -U\Gamma^{-1}$. When partitioned commensurately with the partitions of Y and X given above, this becomes

$$[y \; Y_1 \; Y_2] = [X_1 \; X_2] \begin{bmatrix} \Pi_{10} & \Pi_{11} & \Pi_{12} \\ \Pi_{20} & \Pi_{21} & \Pi_{22} \end{bmatrix} + [v \; V_1 \; V_2] \,. \tag{5}$$

Thus, we can write

$$Y_1 = X_1 \Pi_{11} + X_2 \Pi_{21} + V_1 \,, \tag{6}$$

where Π_{11} is $k_1 \times (g_1 - 1)$ and Π_{21} is $k_2 \times (g_1 - 1)$. The assumption that (3) is identifiable is equivalent to the condition, known as the rank condition, that this latter matrix Π_{21} has full rank $g_1 - 1 \leq k_2$. The relevance of this condition is simply seen by noting that, for the columns of $\Pi \equiv -B\Gamma^{-1}$ corresponding to the given equation, one has $\Pi\gamma = -\beta$, or, partitioned with the appropriate zero restrictions for the omitted variates Y_2 and X_2,

$$\begin{bmatrix} \Pi_{10} & \Pi_{11} & \Pi_{12} \\ \Pi_{20} & \Pi_{21} & \Pi_{22} \end{bmatrix} \begin{pmatrix} -1 \\ \gamma_1 \\ 0 \end{pmatrix} = -\begin{pmatrix} \beta_1 \\ 0 \end{pmatrix} . \tag{7}$$

This gives two equations

$$-\Pi_{10} + \Pi_{11} \gamma_1 = -\beta_1$$

$$-\Pi_{20} + \Pi_{21} \gamma_1 = 0 \,. \tag{8}$$

Clearly, γ_1 is uniquely derivable from the second equation, which is the essence of identifiability, if and only if Π_{21} has full rank. β_1 is then derivable from the first equation.

So, substituting (6) into (3) and gathering terms gives

$$y = X_1(\Pi_{11}\gamma_1 + \beta_1) + X_2\Pi_{21}\gamma_1 + \xi$$

$$= X_1\delta_1 + \tilde{X}_2\gamma_1 + \xi\,, \tag{9}$$

where $\xi = u + V_1\gamma_1 \equiv v$, $\delta_1 \equiv \Pi_{11}\gamma_1 + \beta_1$, and $\tilde{X}_2 \equiv X_2\Pi_{21}$. In the form of the last equality, we see that all of the identifying information — which is contained in the excluded exogenous variates X_2 — is introduced through $\tilde{X}_2 \equiv X_2\Pi_{21}$, which is directly associated with γ_1. And it is here that we can see the relation between true identifiability and the degree of effective identification due to data weaknesses. Clearly, estimation of γ_1 in (9) cannot take place if $\tilde{X}_2$ is of less than full rank (or, more generally, if $\tilde{X}_2$ is ill-conditioned). And $\tilde{X}_2$ will be of less than full rank if either Π_{21} is rank deficient or X_2 is. The rank deficiency of Π_{21} is directly associated with identifiability, for (3) is identifiable if and only if Π_{21} is of full rank, a condition that has nothing to do with the particular data set $[Y\ X]$ available for estimation. However, $\tilde{X}_2 \equiv X_2\Pi_{21}$ will also be of less than full rank (or ill-conditioned) if X_2 is, a condition that is clearly directly related only to the particular data available for estimation and having nothing directly to do with the rank of Π_{21}. Thus, both unidentifiability and the ill-conditioning of X_2 have the same effect in rendering (3) inestimable.

That γ_1 should be the focus of the identifying information is quite understandable, since identification really has only to do with the ability to determine the coefficients of the included endogenous variables. In the discussion surrounding (8) above, we saw that identifiability hinges first on the ability to solve for γ_1 uniquely from the second equation. The solution for β_1 then follows directly from the first equation. From the point of view of estimation this can be seen by noting that, once estimates $\hat{\gamma}_1$ are obtained for γ_1, estimation of β_1 can proceed by least squares of (3) in the form of $y - Y_1\hat{\gamma}_1 = X_1\beta_1 + u$.[4]

Given that the equation is identifiable, then, it is clear as a practical matter that identification will be effective only as long as the $\tilde{X}_2$ are not weak data, that is, as long as the columns of $\tilde{X}_2$ are not short-data or involved in strong collinear relations, either among themselves or with the columns of X_1. But, as is clear from Belsley (1982), this is precisely the condition that the estimator of γ_1 possess adequate signal-to-noise. Thus, if we knew Π_{21} (and hence $\tilde{X}_2$), equation (9) would be amenable to estimation by least squares, and we could apply the test for adequate signal-to-noise given in Belsley (1982) directly to the least-squares estimate $\hat{\gamma}_1$ of γ_1 in order to provide a test for effective identification. Remember, the issue here is not identifiability; the equation (3) is assumed to be identifiable. The issue is whether there is adequate novel information in the excluded exogenous variates to be able effectively to exploit this identifiability.

Unfortunately, we do not typically know Π_{21}, but fortunately this suggested test for effective identification can take place without knowing Π_{21}. Rewrite (9) as

[4] Compare this with the Least Generalized-Variance Ratio Estimator (LGVRE) as derived in Goldberger (1964, p. 338).

$$y = X_1\,\delta_1 + X_2\,\delta_2 + \xi\ , \tag{10}$$

where $\delta_2 \equiv \Pi_{21}\,\gamma_1$. This equation is seen to be that from the reduced form (5) corresponding to the variate y. Under the identifying restrictions, however, we see that the least-squares estimator $\hat{\delta}_2$ of δ_2 must be degenerately distributed by a degree exactly equal to the degree of overidentification $k_2 - g_1 + 1$, since $E\hat{\delta}_2 = \delta_2 \equiv \Pi_{21}\,\gamma_1$. And therefore

$$V(\hat{\delta}_2) = \Pi_{21}\,V(\hat{\gamma}_1)\,\Pi_{21}^T\ , \tag{11}$$

where $V(\hat{\gamma}_1)$ is the variance-covariance matrix of the least-squares estimator $\hat{\gamma}_1$ of γ_1 in (9), and where $V(\hat{\delta}_2)$ is the $k_2 \times k_2$ variance-covariance matrix of $\hat{\delta}_2$ whose rank is $g_1 - 1 \leq k_2$ (since, we recall, that the identifiability of (3) assumes Π_{21} to have full rank equal to $g_1 - 1$).

Because of (11), we see that the signal-to-noise τ^2 for $\hat{\gamma}_1$, for whose adequacy we wish to test, is exactly that for $\hat{\delta}_2$, that is

$$\begin{aligned} \tau^2 &\equiv \gamma_1^T\,V^{-1}(\hat{\gamma}_1)\,\gamma_1 \\ &= \gamma_1^T\,\Pi_{21}^T\,V^+(\hat{\delta}_2)\,\Pi_{21}\,\gamma_1 \\ &= \delta_2^T\,V^+(\hat{\delta}_2)\,\delta_2\ , \end{aligned} \tag{12}$$

where $V^+(\hat{\delta}_2)$ is the generalized inverse of $V(\hat{\delta}_2)$ from (11). The first equality in (12) is obtained as follows:

LEMMA 1 Let P be an $m \times n$ matrix of full rank $n \leq m$, then $P^+P = I$.

Proof From the singular-value decomposition, $P = UDV^T$ and $P^+ = VD^+U^T = VD^{-1}U^T$. Hence $P^+P = VD^{-1}U^TUDV^T = I$.

LEMMA 2 Let $A = PBP^T$, where P is as in Lemma 1 and B is nonsingular, then $A^+ = P^{+T}B^{-1}P^+$ and $B^{-1} = P^TA^+P$.

Proof We must verify that $A^+ \equiv P^{+T}B^{-1}P^+$ satisfies the four conditions that define the generalized inverse (Theil, 1972): (i) $AA^+A = A$, (ii) $A^+AA^+ = A^+$, (iii) $AA^+ = (AA^+)^T$, and (iv) $A^+A = (A^+A)^T$.

i) $(PBP^T)\,(P^{+T}B^{-1}P^+)\,(PBP^T) = PBB^{-1}BP^T = PBP^T$, since, by virtue of Lemma 1, $P^TP^{+T} = (P^+P)^T = I$,

ii) $(P^{+T}B^{-1}P^{+})\,(PBP^{T})\,(P^{+T}B^{-1}P^{+}) =$
$= P^{+T}B^{-1}BB^{-1}P^{+} = P^{+T}B^{-1}P^{+}$,

iii) $(PBP^{T})\,(P^{+T}B^{-1}P^{+}) = PP^{+}$ is symmetric by virtue of P^{+}'s being a generalized inverse, and

iv) $(P^{+T}B^{-1}P^{+})\,(PBP^{T}) = P^{+T}P^{T} = (PP^{+})^{T}$ is symmetric for the same reason. □

Thus, the desired measure of signal-to-noise that is unattainable in the form of the first equality in (12) — because this requires knowing $\tilde{X}_2$ and, hence, Π_{21} to conduct the least-squares estimation of (9) — is attainable in the form of the final equality, which needs only to estimate (10) by least squares. So we may test for the adequacy of the signal-to-noise for γ_1, and hence the degree of effective identification, by testing for the adequacy of the signal-to-noise for δ_2 from the least-squares estimation of an appropriately partitioned reduced-form equation (10).

There will be one minor modification in the direct application of the techniques of Belsley (1982) in this context. Because the rank of $V^{+}(\hat{\delta}_2)$ is $g_1 - 1 \leq k_2$, the k_2-vector $\hat{\delta}_2$ will generally be degenerately distributed, and hence we have

$$(\hat{\delta}_2 - \delta_2)^T V^{+}(\hat{\delta}_2)\,(\hat{\delta}_2 - \delta_2) \overset{A}{\sim} \chi^2_{g_1-1}\,, \tag{13}$$

a chi-squared distribution with $g_1 - 1$ degrees of freedom. Hence the appropriate threshold of adequacy from equation (19) of Belsley (1982) will be

$$\tau^2_{\phi} \equiv {}_{\phi}\chi^2_{g_1-1}\,, \tag{14}$$

rather than ${}_{\phi}\chi^2_{k_2}$, the difference being the degree of overidentification.

Conclusions

The test for adequate signal-to-noise given in Belsley (1982) provides the basis for a measure of the degree of effective identification. The higher the ϕ for which it is possible to accept adequate signal-to-noise (the closer ϕ is to 1), the higher the degree of effective identification. Furthermore, tests for adequate signal-to-noise for subsets of γ_1 (subsets of δ_1) can be interpreted as tests for those coefficients for which there is effective identification versus those for which there is not. This information is useful in determining where additional identifying information, say, through the introduction of prior information, would be most effective in correcting weak identification through zero restrictions. In a related vein, those excluded exogenous variates corresponding to elements of δ_2 that display the greatest signal-to-noise are clearly those that are the best instruments to employ in instrumental-variable estimation, say in the case of an undersized sample, when two-stage least-squares cannot proceed.

Bibliography

Belsley, D.A.: 1982, 'Assessing the Presence of Harmful Collinearity and Other Forms of Weak Data through a Test for Signal-to-Noise', *Journal of Econometrics* **20**, 211–253.

Goldberger, A.S.: 1964, *Econometric Theory*, New York: John Wiley and Sons.

Johnston, J.: 1984, *Econometric Methods*, 3rd Edition, New York: McGraw-Hill.

Malinvaud, E.: 1970, *Statistical Methods of Econometrics*, 2nd revised edition, Amsterdam: North-Holland.

Sims, C.A.: 1980, 'Macroeconomics and Reality', *Econometrica* **48**, 1–48.

Theil, H.: 1971, *Principles of Econometrics*, New York: John Wiley and Sons.

PART TWO

MODEL STIMULATION AND OPTIMIZATION

A SPLITTING EQUILIBRATION ALGORITHM FOR THE COMPUTATION OF LARGE-SCALE CONSTRAINED MATRIX PROBLEMS: THEORETICAL ANALYSIS AND APPLICATIONS

ANNA NAGURNEY

Department of General Business and Finance, School of Management

and

ALEXANDER EYDELAND

Department of Mathematics and Statistics, University of Massachusetts, Amherst, Massachusetts, 01003, U.S.A.

Abstract. In this paper we introduce a general parallelizable computational method for solving a wide spectrum of constrained matrix problems. The constrained matrix problem is a core problem in numerous applications in economics. These include the estimation of input/output tables, trade tables, and social/national accounts, and the projection of migration flows over space and time. The constrained matrix problem, so named by Bacharach, is to compute the best possible estimate X of an unknown matrix, given some information to constrain the solution set, and requiring either that the matrix X be a minimum distance from a given matrix, or that X be a functional form of another matrix. In real-world applications, the matrix X is often very large (several hundred to several thousand rows and columns), with the resulting constrained matrix problem larger still (with the number of variables on the order of the square of the number of rows/columns; typically, in the hundreds of thousands to millions). In the classical setting, the row and column totals are known and fixed, and the individual entries nonnegative. However, in certain applications, the row and column totals need not be known a priori, but must be estimated, as well. Furthermore, additional objective and subjective inputs are often incorporated within the model to better represent the application at hand. It is the solution of this broad class of large-scale constrained matrix problems in a timely fashion that we address in this paper.

The constrained matrix problem has become a standard modelling tool among researchers and practitioners in economics. Therefore, the need for a unifying, robust, and efficient computational procedure for solving constrained matrix problems is of importance. Here we introduce an algorithm, the splitting equilibration algorithm, for computing the entire class of constrained matrix problems. This algorithm is not only theoretically justified, but also fully exploits both the underlying structure of these large-scale problems and the advantages offered by state-of-the-art computer architectures, while simultaneously enhancing the modelling flexibility.

In particular, we utilize some recent results from variational inequality theory, to construct a splitting equilibration algorithm which splits the spectrum of constrained matrix problems into series of row/column equilibrium subproblems. Each such constructed subproblem, due to its special structure, can, in turn, be solved simultaneously via exact equilibration in closed form. Thus each subproblem can be allocated to a distinct processor.

We also present numerical results when the splitting equilibration algorithm is implemented in a serial, and then in a parallel environment. The algorithm is tested against another much-cited algorithm and applied to input/output tables, social accounting matrices, and migration tables. The computational results illustrate the efficacy of this approach.

H.M. Amman et al. (eds), Computational Economics in Econometrics, 65–105.

1. Introduction

In this paper we introduce a general parallelizable computational method for solving a wide spectrum of *Constrained Matrix* problems. The *Constrained Matrix* problem is a core problem in numerous applications. These include the estimation of input-output tables, trade tables, and social/national accounts, the projection of migration flows over space and time, the treatment of census data, the analysis of political voting patterns, and the estimation of contingency tables in statistics. The constrained matrix problem (see Figure 1), so named by Bacharach (1970), is to compute the best possible estimate of an unknown matrix, given some information to constrain the solution set, and requiring either that the matrix be a minimum distance from the given matrix, or that it be a functional form of another known matrix. In real-world applications, the matrix is often very large (several hundred to several thousand rows and columns), with the resulting constrained matrix problem larger still (with the number of variables on the order of the square of the number of rows/columns; typically, in the hundreds of thousands to millions). In the classical setting, the row totals and the column totals are known and fixed, and the individual matrix elements nonnegative. However, in certain applications, the row and column totals need not be known a priori, but must be estimated, as well. Furthermore, additional objective and subjective inputs are often incorporated within the model to better represent the application being studied. It is the solution of this broad class of large-scale constrained matrix problems in a timely fashion that we address in this paper.

The constrained matrix approach has become a standard modelling tool among researchers and practitioners. Therefore, the need for a unifying, robust, and efficient computational procedure for solving constrained matrix problems is of importance. In this paper we introduce an algorithm, which we call the *Splitting Equilibration Algorithm*, for computing solutions to the entire class of constrained matrix problems. The algorithm splits a large-scale constrained matrix problem into series of row (supply)/column (demand) subproblems. Each such constructed subproblem, due to its special structure, can, in turn, be solved simultaneously via *exact equilibration* in closed form on a distinct processor. This algorithm is not only theoretically justified, but also fully exploits the advantages offered by the state-of-the-art computer architectures, while simultaneously enhancing the modelling flexibility.

Our computational procedure is motivated, in part, by the problem at hand - the 'equilibration' of matrices (cf. Van der Sluis (1969)), and by the problem's connection with spatial price equilibrium problems (Enke (1951), Samuelson (1952), and Takayama and Judge (1971)). Indeed, although as early as 1951, Stone recognized that the same methodology should be applied to the computation of both spatial price equilibrium problems and constrained matrix problems, the computational state of the art at the time precluded such an investigation. Furthermore, although the RAS method, which dates to Deming and Stephan (1940), is currently the most

$$\begin{pmatrix} x^0_{11} & x^0_{12} & \cdots & x^0_{1n} \\ x^0_{21} & x^0_{22} & \cdots & x^0_{2n} \\ \vdots & \vdots & \ddots & \vdots \\ x^0_{m1} & x^0_{m2} & \cdots & x^0_{mn} \end{pmatrix} \Longrightarrow \begin{matrix} s_1 \\ s_2 \\ \vdots \\ s_m \end{matrix} \begin{pmatrix} x_{11} & x_{12} & \cdots & x_{1n} \\ x_{21} & x_{22} & \cdots & x_{2n} \\ \vdots & \vdots & \ddots & \vdots \\ x_{m1} & x_{m2} & \cdots & x_{mn} \end{pmatrix}$$
$$\begin{matrix} d_1 & d_2 & \cdots & d_n \end{matrix}$$

Initial Matrix X^0 | Matrix Estimate X

Fig. 1. The constrained matrix problem.

widely applied computational method in practice for the solution of the constrained matrix problem, its limitations include the use of a highly specific set of constraints and objective function and its nonconvergence in certain applications (see, e.g., Mohr, Crown, and Polenske (1987)).

The paper is organized as follows: in Section 2 we present the formulation of the general quadratic constrained matrix problem, specializing it, firstly, to the particular constrained matrix problem arising in the estimation of social accounting matrices in which the row and column totals must 'balance', and, secondly, to the case of known row and column totals. We also relate the general formulation to many currently used models.

In Section 3 we develop the splitting equilibration algorithm and provide a theoretical analysis of the algorithm, including convergence results.

In Sections 4 and 5 we empirically investigate the computational performance of the splitting equilibration algorithm on the largest quadratic constrained matrix problems reported to date using the IBM 3090-600E at the Cornell National Supercomputer Facility. Our goals include: (1) to compare the relative efficiency of the splitting equilibration algorithm to both our earlier equilibration algorithm (RC) (Nagurney, Kim, and Robinson (1990)), and the much-cited Bachem and Korte (1978) algorithm; (2) to investigate the efficiency of the new equilibration approach on the spectrum of very large constrained matrix problems, both diagonal and general; and (3) to investigate the speedups obtained with parallelization of the splitting equilibration algorithm for both diagonal and general large-scale problems.

2. Formulation of the General Constrained Matrix Problem

In this Section we present a general quadratic constrained matrix problem, describe several applications, and highlight the special cases. We also identify its relationship to classical spatial price equilibrium problems.

The problem will be formulated as a minimization of the weighted squared sums of the deviations. We denote the given $m \times n$ matrix by $X^0 = (x^0_{ij'})$, and the matrix estimate by $X = (x_{ij'})$. Let s_i^0 denote the row i total, and s_i the estimate of the row i total. Let $d_{j'}^0$ denote the column j' total, and $d_{j'}$ the estimate of the column j' total. Let the $mn \times mn$ matrix $G = (\gamma_{ij'kl'})$ denote the imposed weight matrix for the mixed variable terms $(x_{ij'} - x_{ij'}^0) \cdot (x_{kl'} - x_{kl'}^0)$ and assume the matrix G to be stictly positive definite. Let the $m \times m$ matrix $A = (\alpha_{ik})$ denote the imposed weight matrix for the mixed variable terms $(s_i - s_i^0) \cdot (s_k - s_k^0)$ and let the $n \times n$ matrix $B = (\beta_{j'l'})$ denote the imposed weight matrix for the mixed variable terms $(d_{j'} - d_{j'}^0) \cdot (d_{l'} - d_{l'}^0)$. Assume that the matrices A and B are also strictly positive definite.

Then the general constrained matrix problem may be written as follows:

$$\text{Minimize} \quad \left[\sum_{i=1}^{m}\sum_{k=1}^{m} \alpha_{ik}(s_i - s_i^0) \cdot (s_k - s_k^0)\right] +$$

$$+ \left[\sum_{i=1}^{m}\sum_{j'=1}^{n}\sum_{k=1}^{m}\sum_{l'=1}^{n} \gamma_{ij'kl'}(x_{ij'} - x_{ij'}^0) \cdot (x_{kl'} - x_{kl'}^0)\right] +$$

$$+ \left[\sum_{j'=1}^{n}\sum_{l'=1}^{n} \beta_{j'l'}(d_{j'} - d_{j'}^0) \cdot (d_{l'} - d_{l'}^0)\right] \tag{1}$$

subject to:

$$\sum_{j'=1}^{n} x_{ij'} = s_i, \quad i = 1, \ldots, m \tag{2}$$

$$\sum_{i=1}^{m} x_{ij'} = d_{j'}, \quad j' = 1, \ldots, n \tag{3}$$

and the inequality constraints:

$$x_{ij'} \geq 0, \quad \text{for all} \quad i, j' \tag{4}$$

where the objective function represents the weighted squared sums of the deviations.

The objective function (1) permits the utilization of mixed-variable weight terms. An example of possibly fully dense A, B, and G matrices are the inverses of the respective variance-covariance matrices (see, e.g., Mosteller and Tukey (1977), Judge and Yancey (1986)). Other examples may arise when the matrices A, B, and G include subjective weights based on the expert knowledge of planners. We note that under the assumption of positive definiteness, the solution to (1), subject to (2) through (4) is unique.

We now describe, as an illustration, the above constrained matrix problem in the context of input/output (I/O) matrices.

An I/O table is constructed from data obtained for a specific economic entity, be it a country, region, state, etc. The economic activity is divided into a number of producing sectors (or industries), with the exchanges of goods between sectors consisting of sales and purchases of goods. Sectors may be general industrial categories (e.g., the aluminun industry), or smaller subdivisions (such as aluminum siding), even larger industrial groups (such as mining ore). The rows of the I/O matrix denote the origin sectors, i.e., the sellers, whereas the columns denote the destination sectors, i.e., the purchasers. The data required to form the I/O table are the flows of the products from each of the producing sectors to the consuming sectors. The data are obtained for a particular time period. The column data represent each sector's inputs while the row data represents each sector's output; thus, the reason for the nomenclature 'input/output table'. For an overview of input/output tables, including applications to regional economics, see Polenske (1980) and Miller and Blair (1985).

The constrained matrix problem arises in the context of I/O tables, when one has a base I/O table for a given time period, typically, a year, and one wishes to update the table to another time period. Often one is not certain of the row and column totals at the new time period. For example, usually, one is able to estimate row and column totals only under certain simplifying assumptions. Furthermore, in countries or regions with a poor information base, the stated row and column totals may simply reflect informed conjectures (see Harrigan and Buchanan (1984) and the references therein). The above formulation takes into account such knowledge gaps directly, thus permitting modelling flexibility.

In the diagonal case, where $\alpha_{ik} = 0$, for $k \neq i$, $\gamma_{ij'kl'} = 0$, for $kl' \neq ij'$, and $\beta_{j'l'} = 0$, for $l' \neq j'$, the objective function (1) simplifies to:

$$\text{Minimize} \left[\sum_{i=1}^{m} \alpha_i (s_i - s_i^0)^2 + \sum_{i=1}^{m} \sum_{j'=1}^{n} \gamma_{ij'} (x_{ij'} - x_{ij'}^0)^2 + \sum_{j'=1}^{n} \beta_{j'} (d_{j'} - d_{j'}^0)^2 \right] \tag{5}$$

subject to the constraints (2) through (4).

The choice of weights is also flexible in this formulation. When the weights in (5) are all equal to one, the problem becomes a constrained least squares problem, and when $\alpha_i = \frac{1}{s_i^0}$, $\beta_{j'} = \frac{1}{d_{j'}^0}$, and $\gamma_{ij'} = \frac{1}{x_{ij'}^0}$, for all rows i and columns j', the objective function is the well-known chi-square. Other possible weights, include $\alpha_i = \frac{1}{s_i^{0\,1/2}}$, $\beta_{j'} = \frac{1}{d_{j'}^{0\,1/2}}$, and $\gamma_{ij'} = \frac{1}{x_{ij'}^{0\,1/2}}$; or a mixed weighting scheme. Of course, when the inverse of the variance-covariance matrix is diagonal, the objective function (5) can also be used.

This diagonal model has been studied by Nagurney (1989) who identified its isomorphism with classical, separable spatial price equilibrium problems and proposed an equilibration algorithm (albeit serial) for its solution. This model is a special case of a constrained matrix problem applied to I/O estimation by Harrigan and Buchanan (1984), who considered interval constraints, rather than equality constraints (2) and (3).

We now consider a special case of the above general quadratic model which arises in the estimation of social/national accounts.

A social accounting matrix (SAM) is a general equilibrium data system, consisting of a series of accounts in the economy of a nation. A SAM is comprised of n rows and n columns, with any particular account, i, represented by row i and column i. The rows represent the receipts of the accounts, the columns the expenditures of the accounts, and the individual matrix entries the transactions in the economy. SAM's have been widely used for policy analyses in developing countries (see, e.g., Pyatt and Round (1985)).

The fact that data used in the construction of SAM's often comes from disparate sources and the need to resolve various inconsistencies motivate the use of the constrained matrix problem in this application. In particular, any specific account represented by the corresponding row and column total must be balanced, that is, the receipts from the accounts must equal the expenditures. This 'definitional' constraint that each row and column must balance makes the SAM estimation problem unique within the domain of constrained matrix problems. In the case when the totals are not known a priori, this gives rise to the following special case of the above general constrained matrix problem:

$$\text{Minimize} \left[\sum_{i=1}^{n}\sum_{k=1}^{n} \alpha_{ik}(s_i - s_i^0)\cdot(s_k - s_k^0)\right] +$$

$$+\left[\sum_{i=1}^{n}\sum_{j'=1}^{n}\sum_{k=1}^{n}\sum_{l'=1}^{n} \gamma_{ij'kl'}(x_{ij'} - x_{ij'}^0)\cdot(x_{kl'} - x_{kl'}^0)\right] \tag{6}$$

subject to inequality constraints (4) and:

$$\sum_{j'=1}^{n} x_{ij'} = s_i, \quad i = 1,\dots,n \tag{7}$$

$$\sum_{i=1}^{n} x_{ij'} = s_{j'}, \quad j' = 1,\dots,n. \tag{8}$$

Stone (1962), Byron (1978), and Van der Ploeg (1982), (1988) considered (6), subject to (7) and (8), but without the inequality constraints (4).

In the special diagonal case, where $\alpha_{ik} = 0$, for all $k \neq i$, and $\gamma_{ij'kl'} = 0$, for all $kl' \neq ij'$, the 'diagonal' objective function for the SAM estimation problem becomes:

$$\text{Minimize} \left[\sum_{i=1}^{n} \alpha_i (s_i - s_i^0)^2 + \sum_{i=1}^{n}\sum_{j'=1}^{n} \gamma_{ij'}(x_{ij'} - x_{ij'}^0)^2\right] \tag{9}$$

subject to the above constraints.

Lastly, in the case where the row and column totals are known with certainty, i.e., $s_i = s_i^0$, for all i, and $d_{j'} = d_{j'}^0$, for all j', the above general quadratic model (1), (2), (3), and (4), collapses to the quadratic constrained matrix problem with fixed row and column totals formulated in Nagurney and Robinson (1989) and solved in Nagurney, Kim, and Robinson (1990) via RC equilibration, whose performance will be compared to the new splitting equilibration algorithm in the subsequent computational sections.

In particular, the objective function in this case simplifies to:

$$\text{Minimize} \sum_{i=1}^{m} \sum_{j'=1}^{n} \sum_{k=1}^{m} \sum_{l'=1}^{n} \gamma_{ij'kl'}(x_{ij'} - x_{ij'}^0) \cdot (x_{kl'} - x_{kl'}^0) \tag{10}$$

with the inequality constraints (4) and:

$$\sum_{j'=1}^{n} x_{ij'} = s_i^0, \quad i = 1, \ldots, m \tag{11}$$

$$\sum_{i=1}^{m} x_{ij'} = d_{j'}^0, \quad j' = 1, \ldots, n. \tag{12}$$

This model may also be applied to the estimation of input/output tables, and social/national accounting matrices, provided that the row and column totals are known with certainty, as well as to the estimation of migration flows. Migration tables are used to study growth patterns in distinct locations and to predict the needs for public services and housing. In migration tables rows represent the origin locations and columns the destination locations. The matrix entries represent the population flows between the origins and destinations. Row and column totals may be available for such applications, for example, when the totals have been obtained for a time period in the past, and the matrix entries which yield those totals are needed, given matrix entries for an earlier time period.

In the much-studied diagonal constrained matrix problem with fixed row and column totals, where the $\gamma_{ij'kl'} = 0$, for $kl' \neq ij'$, the objective function (10) further simplifies to

$$\text{Minimize} \sum_{i=1}^{m} \sum_{j'=1}^{n} \gamma_{ij'}(x_{ij'} - x_{ij'}^0)^2 \tag{13}$$

with constraints (11), (12), and (4).

Deming and Stephan (1940) considered (13) with $\gamma_{ij'} = \frac{1}{x_{ij'}^0}$, subject to constraints (11) and (12), whereas Friedlander (1961) considered the case where $G = I$. Bachem and Korte (1978) treated (13) with all of the constraints. Cottle, et al. (1986), on the other hand, studied Friedlander's problem with the additional constraints (4). Ohuchi and Kaji (1984) studied the Bachem and Korte problem with upper and lower bounds. Klincewicz (1989) also studied the above diagonal model.

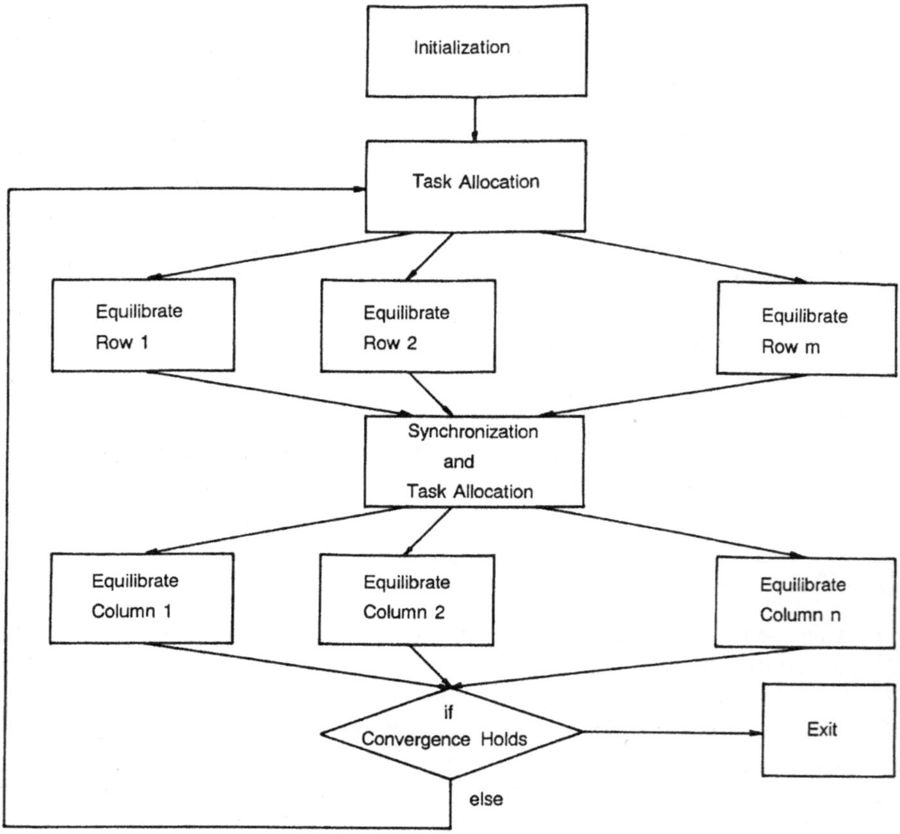

Fig. 2. Flowchart of SEA – the diagonal case.

3. The Splitting Equilibration Algorithm

In this Section we introduce the splitting equilibration algorithm (SEA) for the computation of the general quadratic constrained matrix problem and its variants.

As mentioned in the Introduction, the algorithm is motivated both by the tremendous need for a unified and robust procedure which can solve the wide class of constrained matrix problems outlined in the preceding section and by the desire to exploit the underlying structure of these problems and the advantages of state-of-the-art computer architectures to allow the solution of problems of a scale larger than heretofore has been practical.

The algorithm constructs a series of diagonal problems of the form (5), subject to the constraints (2), (3), and (4). Each of the constructed diagonal problems, in turn, is then 'split' to handle the row constraints (2) and then the column constraints (3). Each such row (supply) or column (demand) subproblem, because of the

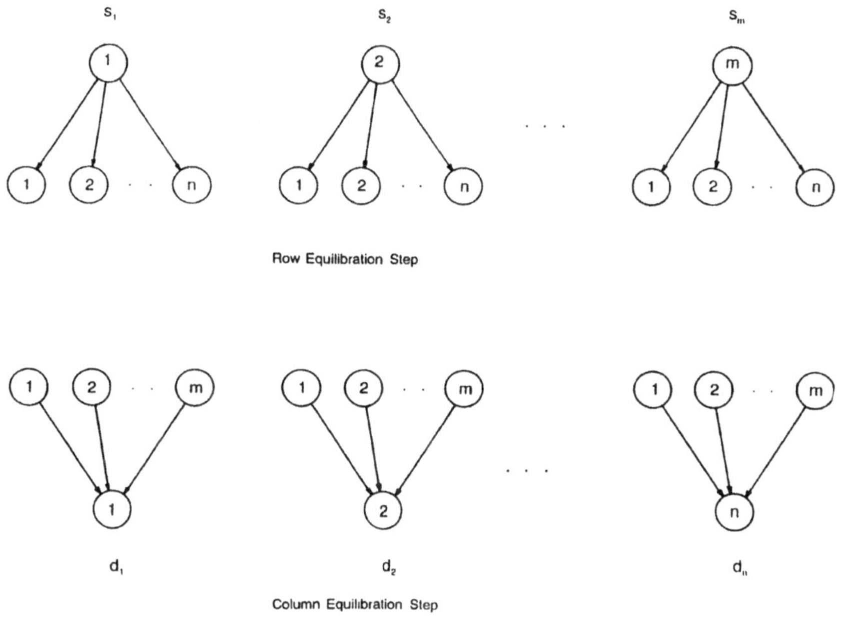

Fig. 3. Network structure of row and column equilibrium subproblems.

separability of the diagonal problem and the splitting of the constraints, can be solved simultaneously, i.e., in parallel, on a distinct processor. In a similar manner, the algorithm diagonalizes the SAM objective function (6), and the objective function (10) in the problem with fixed row and column totals, and then splits the row and column constraints (7), (8), and (11), (12), respectively, inducing, again, row and column equilibrium subproblems which can be solved exactly in closed form, and in parallel. Of course, in the diagonal constrained matrix problems with objective functions given, respectively, by (5), (9), and (13), no such diagonalization is required, before the splitting occurs.

We first present the splitting equilibration algorithm for the diagonal problems and then for the general ones. The general problems are solved iteratively using the appropriate diagonal procedure outlined immediately below. A graphical depiction of SEA for the diagonal problems is given in Figure 2. The special structure of the row and column equilibrium subproblems is then highlighted in Figure 3. A graphical depiction of SEA for general problems is given in Figure 4.

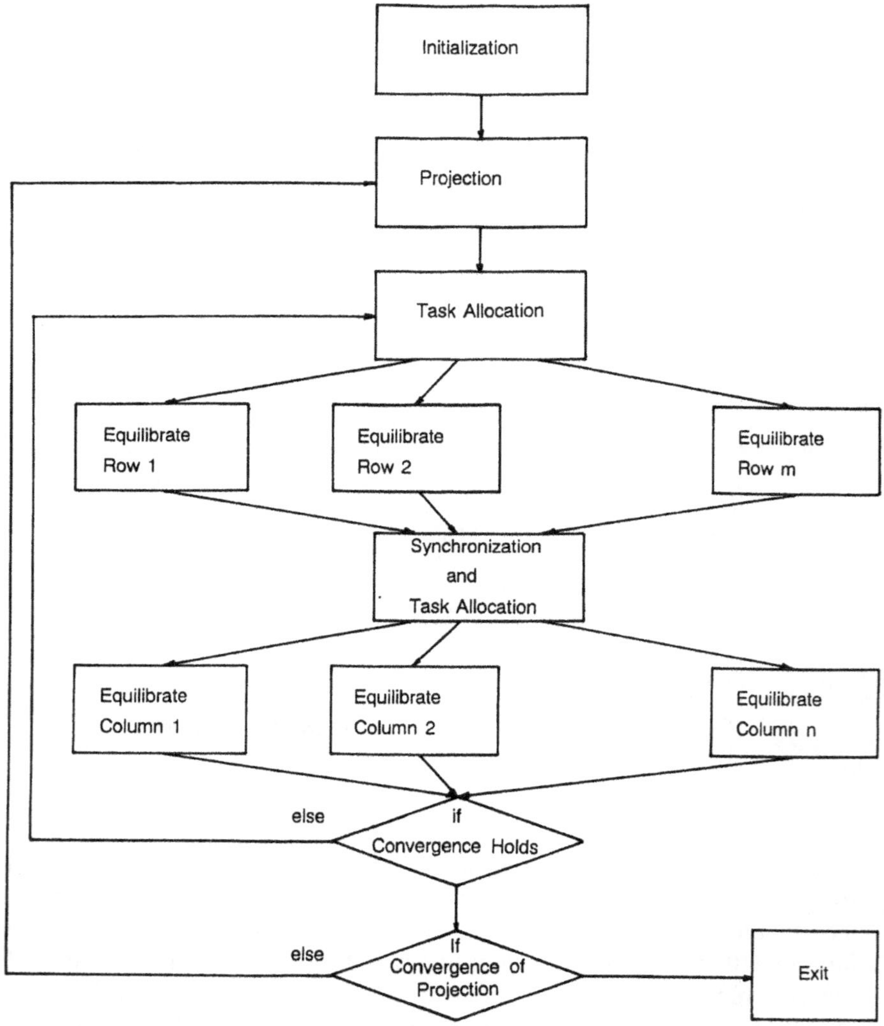

Fig. 4. Flowchart of SEA – the general case.

3.1. The Splitting Equilibration Algorithm for Diagonal Problems

In this Section we present the splitting equilibration algorithm for the diagonal problems formulated in Section 2. We first focus on the diagonal problem with unknown row and column totals and objective function (1). We then describe SEA for the SAM estimation problem with objective function (9) and conclude with the constrained matrix problem with fixed row and column totals and objective function (13). We conclude with a unified interpretation of the algorithm as a dual method and provide theoretical results.

3.1.1. SEA for the Constrained Matrix Problem with Unknown Row and Column Totals

This algorithm computes a solution to problem (5), i.e.,

$$\text{Minimize } \Theta_1(x,s,d) = \sum_{i=1}^{m} \alpha_i (s_i - s_i^0)^2$$

$$+\sum_{i=1}^{m}\sum_{j'=1}^{n} \gamma_{ij'}(x_{ij'} - x_{ij'}^0)^2 + \sum_{j'=1}^{n} \beta_{j'}(d_{j'} - d_{j'}^0)^2$$

subject to constraints (2), (3), and (4).

Step 0: Initialization Step
Let $\mu^1 \in R^n = 0$. Set $t = 1$.

Step 1: Row Equilibration
Find $X(\mu^t), S(\mu^t), D(\mu^t)$ such that

$$(X(\mu^t), S(\mu^t), D(\mu^t)) \rightarrow \min_{x,s,d} \Theta_1(x,s,d) - \sum_{j'=1}^{n} \mu_{j'}^t (\sum_{i=1}^{m} x_{ij'} - d_{j'}) \tag{14}$$

subject to:

$$\sum_{j'=1}^{n} x_{ij'} = s_i, \quad i = 1, \ldots, m \tag{15}$$

$$x_{ij'} \geq 0.$$

Compute the corresponding Lagrange multipliers λ_i^{t+1} for this problem, according to $\lambda_i^{t+1} = 2\alpha_i s_i^0$ -$2\alpha_i\, S_i(\mu^t)$, $i = 1, \ldots, m$, and use them in Step 2.

Step 2: Column Equilibration
Find $X(\lambda^{t+1}), S(\lambda^{t+1}), D(\lambda^{t+1})$ such that

$$(X(\lambda^{t+1}), S(\lambda^{t+1}), D(\lambda^{t+1})) \rightarrow \min_{x,s,d} \Theta_1(x,s,d) - \sum_{i=1}^{m} \lambda_i^{t+1} (\sum_{j'=1}^{n} x_{ij'} - s_i) \tag{16}$$

subject to:

$$\sum_{i=1}^{m} x_{ij'} = d_{j'}, \quad j' = 1, \ldots, n \tag{17}$$

$$x_{ij'} \geq 0.$$

Compute the corresponding Lagrange multipliers μ_j^{t+1} for this problem, according to $\mu_{j'}^{t+1} = 2\beta_{j'}\, d_{j'}^0 - 2\beta_{j'}\, D_{j'}\,(\lambda^{t+1})$, $j' = 1, \ldots, n$, and use them in Step 1.

Step 3: Convergence Verification

If $|\sum_{j'} x_{ij'}(\lambda^{t+1}) - S_i(\mu^t)| \leq \epsilon$; for all i, j', terminate; else, set $t = t + 1$, and go to Step 1.

Observe that both the row and column equilibration problems correspond to m and n disjoint network subproblems, respectively, each of which can be allocated to a distinct processor and solved, respectively, using the supply market/demand market exact equilibration algorithms discussed in Eydeland and Nagurney (1989).

We now provide a dual interpretation of the above algorithm which is used in establishing convergence and the theoretical analysis.

We introduce the Lagrangian

$$L_1(x, s, d, \lambda, \mu) = \Theta_1(x, s, d) - \sum_{i=1}^{m} \lambda_i (\sum_{j'=1}^{n} x_{ij'} - s_i) - \sum_{j'=1}^{n} \mu_{j'} (\sum_{i=1}^{m} x_{ij'} - d_{j'}) \tag{18}$$

and

$$\zeta_1(\lambda, \mu) = \min_{x \geq 0, s, d} L_1(x, \lambda, \mu). \tag{19}$$

We also let $\lambda^*, \mu^* \rightarrow \max \zeta_1(\lambda, \mu)$. In (19), the optimal values $X = X(\lambda, \mu)$, $S = S(\lambda, \mu)$, $D = D(\lambda, \mu)$ satisfy the following relations:

$$\frac{\partial L_1}{\partial x_{ij'}}(X, S, D, \lambda, \mu) = 2\gamma_{ij'} X_{ij'} - 2\gamma_{ij'} x_{ij'}^0 - \lambda_i - \mu_{j'} \tag{20}$$

$$\begin{cases} \geq 0, & \text{if} \quad X_{ij'} = 0 \\ = 0, & \text{if} \quad X_{ij'} > 0 \end{cases}$$

$$\frac{\partial L_1}{\partial s_i}(X, S, D, \lambda, \mu) = 2\alpha_i S_i - 2\alpha_i s_i^0 + \lambda_i = 0 \tag{21}$$

$$\frac{\partial L_1}{\partial d_{j'}}(X, S, D, \lambda, \mu) = 2\beta_{j'} D_{j'} - 2\beta_{j'} d_{j'}^0 + \mu_{j'} = 0. \tag{22}$$

Therefore,

$$X_{ij'} = \frac{1}{2\gamma_{ij'}}(2\gamma_{ij'} x_{ij'}^0 + \lambda_i + \mu_{j'})_+ \tag{23a}$$

$$S_i = \frac{1}{2\alpha_i}(2\alpha_i s_i^0 - \lambda_i) \tag{23b}$$

$$D_{j'} = \frac{1}{2\beta_{j'}}(2\beta_{j'} d_{j'}^0 - \mu_{j'}). \tag{23c}$$

Hence,

$$\zeta_1(\lambda,\mu) \equiv L_1(X,S,D,\lambda,\mu)$$

$$= \sum_{ij'} \gamma_{ij'}\Big(\frac{1}{2\gamma_{ij'}}(2\gamma_{ij'}x^0_{ij'} + \lambda_i + \mu_{j'})_+ - x^0_{ij'}\Big)^2$$

$$+ \sum_i \alpha_i\Big(\frac{1}{2\alpha_i}(2\alpha_i s^0_i - \lambda_i) - s^0_i\Big)^2 + \sum_{j'} \beta_{j'}\Big(\frac{1}{2\beta_{j'}}(2\beta_{j'}d^0_{j'} - \mu_{j'}) - d^0_{j'}\Big)^2$$

$$- \sum_i \lambda_i\Big(\sum_{j'} \frac{1}{2\gamma_{ij'}}(2\gamma_{ij'}x^0_{ij'} + \lambda_i + \mu_{j'})_+ - \frac{1}{2\alpha_i}(2\alpha_i s^0_i - \lambda_i)\Big)$$

$$- \sum_{j'} \mu_{j'}\Big(\sum_i \frac{1}{2\gamma_{ij'}}(2\gamma_{ij'}x^0_{ij'} + \lambda_i + \mu_{j'})_+ - \frac{1}{2\beta_{j'}}(2\beta_{j'}d^0_{j'} - \beta_{j'})\Big)$$

$$= - \sum_{i,j'} \frac{1}{4\gamma_{ij'}}(2\gamma_{ij'}x^0_{ij'} + \lambda_i + \mu_{j'})^2_+$$

$$- \sum_i \frac{1}{4\alpha_i}(2\alpha_i s^0_i - \lambda_i)^2 - \sum_{j'} \frac{1}{4\beta_{j'}}(2\beta_{j'}d_{j'} - \mu_{j'})^2$$

$$+ \sum_{ij'} \gamma_{ij'}{x^0_{ij'}}^2 + \sum_i \alpha_i s^{02}_i + \sum_{j'} \beta_{j'} d^{02}_j. \tag{24}$$

The last three terms in (24) are constant and, therefore, can be ignored.

Moreover,

$$\frac{\partial \zeta_1(\lambda,\mu)}{\partial \lambda_i} = - \sum_{j'} \frac{1}{2\gamma_{ij'}}(2\gamma_{ij'}x^0_{ij'} + \lambda_i + \mu_{j'})_+ + \frac{1}{2\alpha_i}(2\alpha_i s^0_i - \lambda_i)$$

$$\equiv - \sum_{j'} X_{ij'}(\lambda,\mu) + S_i(\lambda,\mu). \tag{25}$$

$$\frac{\partial \zeta_1}{\partial \mu_{j'}} = - \sum_i \frac{1}{2\gamma_{ij'}}(2\gamma_{ij'}x^0_{ij'} + \lambda_i + \mu_{j'})_+$$

$$+ \frac{1}{2\beta_{j'}}(2\beta_{j'}d^0_{j'} - \mu_{j'}) \equiv D_{j'}(\lambda,\mu) - \sum_i X_{ij'}(\lambda,\mu). \tag{26}$$

Therefore,

$$\|\nabla \zeta_1(\lambda,\mu)\| \le \epsilon \sim \|\text{Constraints}\| \le \epsilon. \tag{27}$$

We define λ^{t+1}, μ^{t+1} according to

$$\lambda^{t+1} \rightarrow \max_{\lambda} \zeta_1(\lambda, \mu^t) \tag{28a}$$

$$\mu^{t+1} \rightarrow \max_{\mu} \zeta_1(\lambda^{t+1}, \mu). \tag{28b}$$

By the duality principle, step (28a) is equivalent to solving precisely the problem:

$$\text{Minimize} \quad \Theta_1(x, s, d) - \sum_{j'=1}^{n} \mu_{j'}^{t} (\sum_{i=1}^{m} x_{ij'} - d_{j'}) \tag{29a}$$

subject to

$$\sum_{j'=1}^{n} x_{ij'} = s_i, \quad i = 1, \ldots, m$$

$$x_{ij'} \geq 0$$

with λ^{t+1} being the Lagrange multipliers for the above row constraints. Note that if $(X(\mu^t), S(\mu^t), D(\mu^t))$ solves (29a), then we can find λ^{t+1} using the same argument for (23b):

$$\lambda_i^{t+1} = 2\alpha_i s_i^0 - 2\alpha_i S_i(\mu^t). \tag{29b}$$

Observe that (29a–29b) is precisely the Row Equilibration Step 1.

Analogously, the step (28b) is equivalent to:

$$\text{Minimize} \quad \Theta_1(x, s, d) - \sum_{i=1}^{m} \lambda_i^{t+1} (\sum_{j'=1}^{n} x_{ij'} - s_i) \tag{30a}$$

subject to

$$\sum_{i=1}^{m} x_{ij'} = d_{j'}$$

$$x_{ij'} \geq 0$$

with μ^{t+1} being the corresponding Lagrange multipliers. Again, if $(X(\lambda^{t+1}), S(\lambda^{t+1}), D(\lambda^{t+1}))$ is the solution of (30a), then

$$\mu_j^{t+1} = 2\beta_{j'} d_{j'}^0 - 2\beta_{j'} D_{j'}(\lambda^{t+1}). \tag{30b}$$

Thus, (30a–30b) is equivalent to the Column Equilibration Step 2.

In the case of the diagonal SAM constrained matrix problem, we note that $m = n$, and $s_i = d_i$, for all i. The statement of the splitting equilibration algorithm for this constrained matrix problem is presented in the following section.

3.1.2. SEA for the SAM Constrained Matrix Problem

This algorithm computes a solution to problem (9), i.e.,

$$\text{Minimize} \quad \Theta_2(x, s) = \sum_{j'=1}^{n} \alpha_{j'} (s_j - s_{j'}^0)^2 + \sum_{i=1}^{n} \sum_{j'=1}^{n} \gamma_{ij'} (x_{ij'} - x_{ij'}^0)^2 \tag{31}$$

subject to constraints (7), (8), and (4).

Step 0: Initialization Step
Let $\mu^1 \in R^n = 0$. Set $t = 1$.

Step 1: Row Equilibration
Find $X(\mu^t), S(\mu^t)$ such that

$$(X(\mu^t), S(\mu^t)) \rightarrow \min_{x,s} \Theta_2(x,s) - \sum_{j'=1}^{n} \mu_{j'}^t (\sum_{i=1}^{n} x_{ij'} - s_{j'}) \tag{32}$$

subject to:

$$\sum_{j'=1}^{n} x_{ij'} = s_i, \quad i = 1, \ldots, n \tag{33}$$

$$x_{ij'} \geq 0.$$

Compute the corresponding Lagrange multipliers λ_i^{t+1}, by

$$\lambda_i^{t+1} = -2\alpha_{i'} S_i(\mu^t) + 2\alpha_{i'} s_i^0 - \mu_{i'}^t, i = 1, \ldots, n,$$

and use them in Step 2.

Step 2: Column Equilibration
Find $X(\lambda^{t+1}), S(\lambda^{t+1})$ such that

$$(X(\lambda^{t+1}), S(\lambda^{t+1})) \rightarrow \min_{x,s} \Theta_2(x,s) - \sum_{i=1}^{n} \lambda_i^{t+1} (\sum_{j'=1}^{n} x_{ij'} - s_i) \tag{34}$$

subject to:

$$\sum_{i=1}^{m} x_{ij'} = s_{j'}, \quad j' = 1, \ldots, n \tag{35}$$

$$x_{ij'} \geq 0.$$

Compute the corresponding Lagrange multipliers μ^{t+1}, by

$$\mu_{j'}^{t+1} = -2\alpha_{j'} S_{j'}(\mu^t) + 2\alpha_{j'} s_{j'}^0 - \lambda_{j'}^{t+1}, j' = 1, \ldots, n,$$

and use them in Step 1.

Step 3: Convergence Verification
If $| \sum_{j'} X_{ij'}(\lambda^{t+1} - S_i(\mu^t)| \leq \epsilon'$, for all i, terminate; else, set $t = t + 1$, and go to Step 1.

The above row and column equilibration problems have the same special structure as that encountered in the analogous subproblems in Section 3.1.1, and again are amenable to solution via exact equilibration on distinct processors.

We now provide a dual interpretation of the above algorithm.

We introduce the Lagrangian

$$L_2(x,s,\lambda) = \Theta_2(x) - \sum_i \lambda_i(\sum_{j'} x_{ij'} - s_{i'}) - \sum_{j'} \mu_{j'}(\sum_i x_{ij'} - s_{j'}) \tag{36}$$

and

$$\zeta_2(\lambda,\mu) = \min_{x\geq 0,s} L_2(x,s,\lambda,\mu) \tag{37}$$

and we let $\lambda^*, \mu^* \rightarrow \max \zeta_2(\lambda,\mu)$.

Since

$$\frac{\partial L_2}{\partial x_{ij'}} = 2\gamma_{ij'}X_{ij'} - 2\gamma_{ij'}x^0_{ij'} - \lambda_i - \mu_{j'} \tag{38}$$

$$\begin{cases} \geq 0, & \text{if} \quad X_{ij'} = 0 \\ = 0, & \text{if} \quad X_{ij'} > 0 \end{cases}$$

and

$$\frac{\partial L_2}{\partial s_{j'}} = 2\alpha_{j'}S_{j'} - 2\alpha_{j'}s^0_{j'} + \lambda_j + \mu_{j'} = 0, \tag{39}$$

implies that:

$$X_{ij'} = \frac{1}{2\gamma_{ij'}}(2\gamma_{ij'}x^0_{ij'} + \lambda_i + \mu_{j'})_+ \tag{40a}$$

$$S_{j'} = \frac{1}{2\alpha_{j'}}(2\alpha_{j'}s^0_{j'} - \lambda_j - \mu_{j'}), \tag{40b}$$

with $\zeta_2(\lambda,\mu)$ taking the form:

$$\begin{aligned} \zeta_2(\lambda,\mu) = & -\sum_{ij'} \frac{1}{4\gamma_{ij'}}(2\gamma_{ij'}x^0_{ij'} + \lambda_i + \mu_{j'})^2_+ \\ & -\sum_j \frac{1}{4\alpha_{j'}}(2\alpha_{j'}s^0_{j'} - \lambda_j - \mu_{j'})^2 \\ & +\sum_{ij'} \gamma_{ij'}x^0_{ij'} + \sum_{j'} \alpha_{j'}s^0_{j'} \end{aligned} \tag{41}$$

with the last two terms in (41) being fixed.

Moreover, in regards to the gradient $\nabla\zeta_2(\lambda,\mu)$ we observe that

$$\frac{\partial \zeta_2}{\partial \lambda_i} = -\sum_{j'} \frac{1}{2\gamma_{ij'}}(2\gamma_{ij'}x^0_{ij'} + \lambda_i + \mu_{j'})_+ + \frac{1}{2\alpha_{i'}}(2\alpha_{i'}s^0_{i'} - \lambda_i - \mu_{i'}) \tag{42}$$

which implies that:

$$\|\text{Constraints}\| \leq \epsilon \sim \|\nabla\zeta_2(\lambda,\mu)\| \leq \epsilon. \tag{43}$$

SEA for the SAM estimation problem, hence, has the following dual interpretation:

$$\lambda^{t+1} \rightarrow \max_{\lambda} \zeta_2(\lambda, \mu^t) \tag{44a}$$

and

$$\mu^{t+1} \rightarrow \max_{\mu} \zeta_2(\lambda^{t+1}, \mu) \tag{44b}$$

where (44a) correspond to the row equilibration step and (44b) correspond to the column equilibration step.

Finally, we present the statement of the splitting equilibration algorithm for the case of fixed row and column totals, which is equivalent to the diagonal RC algorithm described in Nagurney, Kim, and Robinson (1990), but which will be theoretically analyzed below.

3.1.3. SEA for the Constrained Matrix Problem with Fixed Row and Column Totals

This algorithm computes the solution to problem (13), i.e.,

$$\text{Minimize} \quad \Theta_3(x) = \sum_{i=1}^{m}\sum_{j'=1}^{n} \gamma_{ij'}(x_{ij'} - x_{ij'}^0)^2$$

subject to (2), (3), and (4).

Step 0: Initialization Step
Let $\mu^1 \in R^n = 0$. Set $t = 1$.

Step 1: Row Equilibration
Find $X(\mu^t)$ such that

$$X(\mu^t) \rightarrow \min_{x} \Theta_3(x) - \sum_{j'=1}^{n} \mu_{j'}^t (\sum_{i=1}^{m} x_{ij'} - d_{j'}^0) \tag{45}$$

subject to

$$\sum_{j'=1}^{n} x_{ij'} = s_i^0, \quad i = 1, \ldots, m \tag{46}$$

$$x_{ij'} \geq 0.$$

Compute the corresponding Lagrange multipliers λ^{t+1}.

Step 2: Column Equilibration
Find $X(\lambda^{t+1})$ such that

$$(X(\lambda^{t+1})) \to \min_{x} \Theta_3(x) - \sum_{i=1}^{m} \lambda_i^{t+1} (\sum_{j'=1}^{n} x_{ij'} - d_{j'}^0) \tag{47}$$

subject to

$$\sum_{i=1}^{m} x_{ij'} = d_{j'}^0, \quad j' = 1, \ldots, n \tag{48}$$

$$x_{ij'} \geq 0.$$

Obtain the corresponding Lagrange multipliers μ^{t+1}.

Step 3: Convergence Verification
Same as Step 3 above with $S_i(\mu^t) = s_i^0$, for all i.

The row and column equilibration problems above again have a characteristic network structure where, however, the respective equilibration problems differ from those encountered in 3.1.1 and 3.1.2 in that they are of the 'fixed' type, rather than elastic, in that s_i^0, $i = 1, \ldots, m$ and $d_{j'}^0$, $j' = 1, \ldots, n$ are known and fixed.

The dual interpretation now follows.

We introduce the Lagrangian

$$L_3(x, \lambda, \mu) = \Theta_3(x) - \sum_{i=1}^{m} \lambda_i (\sum_{j'=1}^{n} x_{ij'} - s_i^0) - \sum_{j'=1}^{n} \mu_{j'} (\sum_{i=1}^{m} x_{ij'} - d_{j'}^0) \tag{49}$$

and

$$\zeta_3(\lambda, \mu) = \min_{x \geq 0} L_3(x, \lambda, \mu). \tag{50}$$

By direct computations we obtain that

$$\zeta_3(\lambda, \mu) = -\sum_{ij'} \frac{1}{4\gamma_{ij'}} (2\gamma_{ij'} x_{ij'}^0 + \lambda_i + \mu_{j'})_+^2$$

$$+ \sum_i \lambda_i s_i^0 + \sum_{j'} \mu_{j'} d_{j'}^0 + \sum_{ij'} \gamma_{ij'} x_{ij'}^{0\,2}. \tag{51}$$

Again,

$$\|\zeta_3(\lambda, \mu)\| \leq \epsilon \sim \|\text{Constraints}\| \leq \epsilon. \tag{52}$$

The algorithm now is the same, as the two previous ones, i.e.,

$$\lambda^{t+1} \to \max_{\lambda} \zeta_3(\lambda, \mu^t) \tag{53a}$$

$$\mu^{t+1} \rightarrow \max_{\mu} \zeta_3(\lambda^{t+1}\mu). \tag{53b}$$

Step (53a) is equivalent to solving the problem

$$\text{Minimize}_x \Theta_3(x) - \sum_{j'=1}^{n} \mu_{j'} (\sum_{i=1}^{m} x_{ij'} - d^0_{j'}) \tag{54a}$$

subject to

$$\sum_{j'=1}^{n} x_{ij'} = s^0_i$$

$$x_{ij'} \geq 0,$$

with λ^t being the Lagrange multipliers for this problem. If $X(\mu^t)$ is a solution of (54a) then

$$\frac{\partial \Theta_3}{\partial x_{ij'}}(X(\mu^t)) - \mu^t_{j'} = \lambda^{t+1}_i, \quad \text{for every} \quad i = 1, \ldots, m, \quad j' = 1, \ldots, n. \tag{54b}$$

This expression indicates how to update the λ's.

Step (53b), analogously, is equivalent to

$$\text{Minimize}_x \Theta_3(x) - \sum_{i=1}^{m} \lambda^{t+1}_i (\sum_{j'=1}^{n} x_{ij'} - s^0_i) \tag{55a}$$

subject to

$$\sum_{i=1}^{m} x_{ij'} = d^0_{j'}$$

$$x_{ij'} \geq 0,$$

with μ^{t+1} being the corresponding Lagrange multipliers. Also, we have that

$$\frac{\partial \Theta_3}{\partial x_{ij'}}(X(\lambda^{t+1})) - \lambda^{t+1}_i = \mu^{t+1}_{j'}, \quad \text{for every} \quad i = 1, \ldots, m, \quad j' = 1, \ldots, n, \tag{55b}$$

where $X(\lambda^{t+1})$ is a solution of (55a).

We now provide a summarization and unification of the above algorithms and refer, henceforth, to the method as SEA.

The Algorithm:

$$\lambda^{t+1} \rightarrow \max \zeta_l(\lambda, \mu^t)$$

$$\mu^{t+1} \rightarrow \max \zeta_l(\lambda^{t+1}, \mu)$$

$$l = 1, 2, 3$$

where

$$\zeta_1(\lambda,\mu) = -\sum_i\sum_{j'}\frac{1}{4\gamma_{ij'}}(2\gamma_{ij'}x^0_{ij'}+\lambda_i+\mu_{j'})^2_+ - \sum_i\frac{1}{4\alpha_i}(2\alpha_i s^0_i-\lambda_i)^2$$

$$-\sum_{j'}\frac{1}{4\beta_{j'}}(2\beta_{j'}d^0_{j'}-\mu_{j'})^2+\sum_{ij'}\gamma_{ij'}x^{0\;2}_{ij'}+\sum_i\alpha_i s^{02}_i+\sum_{j'}d^{0\;2}_{j'},$$

$$\zeta_2(\lambda,\mu) = -\sum_i\sum_{j'}\frac{1}{4\gamma_{ij'}}(2\gamma_{ij'}x^0_{ij'}+\lambda_i+\mu_{j'})^2_+ - \sum_{j'}\frac{1}{4\alpha_{j'}}(2\alpha_{j'}s^0_{j'}-\lambda_j-\mu_{j'})^2$$

$$+\sum_{ij'}\gamma_{ij'}x^{0\;2}_{ij'}+\sum_{j'}\alpha_{j'}s^{0\;2}_{j'},$$

$$\zeta_3(\lambda,\mu) = -\sum_{ij'}\frac{1}{4\gamma_{ij'}}(2\gamma_{ij'}x^0_{ij'}+\lambda_i+\mu_{j'})^2_+ + \sum_i\lambda_i s^0_i+\sum_{j'}\mu_{j'}d^0_{j'}+\sum_{ij'}\gamma_{ij'}x^{0\;2}_{ij'}.$$

We now state the proof of convergence using the stopping criterion: $\|\zeta_l(\lambda,\mu)\|\le\epsilon$.

Observe that

$$\zeta_l(\lambda^{t+1},\mu^{t+1})\ge\max_\tau\zeta_l(\lambda^{t+1},\mu^t+\tau d^t) \tag{56}$$

with

$$d^t=\frac{\nabla_\mu\zeta_l(\lambda^{t+1},\mu^t)}{\|\nabla_\mu\zeta_l(\lambda^{t+1},\mu^t)\|}=\frac{\nabla\zeta_l(\lambda^{t+1},\mu^t)}{\|\nabla\zeta_l(\lambda^{t+1},\mu^t)\|}. \tag{57}$$

Let us now estimate $\tau_{\max}$ where τ is the value for which the max in (56) is attained.

Since on the interval $[\mu^t,\mu^t+\tau_{\max}d^t]$ the function $\theta(\tau)$=$\nabla\zeta_l(\lambda^t,\mu^t+\tau d^t)\times d^t$ changes from $\|\nabla\zeta_l(\lambda^{t+1},\mu^t)\|$ to 0 and $M_l\ge|\frac{\partial\theta_l}{\partial\tau}|\ge m_l$, $l=1,2,3$, where

$$m_1=\min\{\min_{ij'}\frac{1}{2\gamma_{ij'}},\quad\min_i\frac{1}{2\alpha_i},\quad\min_{j'}\frac{1}{2\beta_{j'}}\} \tag{58a}$$

$$m_2=\min\{\min_{ij'}\frac{1}{2\gamma_{ij'}},\quad\min_i\frac{1}{2\alpha_i}\} \tag{58b}$$

$$m_3=\min_{ij'}\frac{1}{2\gamma_{ij'}} \tag{58c}$$

and

$$M_1=\max\{\max_{ij'}\frac{1}{2\gamma_{ij'}},\quad\max_i\frac{1}{2\alpha_i},\quad\max_{j'}\frac{1}{2\beta_{j'}}\} \tag{59a}$$

$$M_2=\max\{\max_{ij'}\frac{1}{2\gamma_{ij'}},\quad\max_i\frac{1}{2\alpha_i}\} \tag{59b}$$

$$M_3=\max_{ij'}\frac{1}{2\gamma_{ij'}}, \tag{59c}$$

we can conclude that:

$$\tau_{\max} \geq \frac{1}{M_l} \|\nabla \zeta_l(\lambda^{t+1}, \mu^t)\|. \tag{60}$$

Expanding now $\zeta_l(\lambda^{t+1}, \mu^t + \tau d^t)$ around $\tau = \tau_{\max}$, we get

$$\zeta_l(\lambda^{t+1}, \mu^t) \leq \zeta_l(\lambda^{t+1}, \mu^t + \tau_{\max} d^t) - \frac{\tau_{\max}{}^2}{2} m_l. \tag{61}$$

Hence,

$$\zeta_l(\lambda^{t+1}, \mu^t) \leq \zeta_l(\lambda^{t+1}, \mu^{t+1}) - \frac{m_l}{2M_l^2} \|\nabla \zeta_l(\lambda^{t+1}, \mu^t)\|^2 \tag{62}$$

or

$$\delta^t = \zeta_l(\lambda^{t+1}, \mu^{t+1}) - \zeta_l(\lambda^{t+1}, \mu^t) \geq \frac{m_l}{2M_l^2} \|\nabla \zeta_l(\lambda^{t+1}, \mu^t)\|^2 \geq \frac{m_l}{2M_l^2} \epsilon^2 \tag{63}$$

where ϵ is the stopping criterion.

Thus, our algorithm must stop in no more than

$$T = \frac{\zeta_{\max} - \zeta_l(\lambda^1, \mu^0)}{\frac{m_l}{2M_l^2}} \times \frac{1}{\epsilon^2} \quad \text{steps.} \tag{64}$$

We note that while dual algorithms have been proposed by Cottle, Duvall, and Zikan (1986) and Ohuchi and Kaji (1984) for the fixed model with objective function (13) and constraints (11) and (12), this is the first such unified treatment of both fixed and the more general elastic versions. For other dual methods and associated applications, see Bertsekas and Tsitsiklis (1989). Moreover, our proof of convergence is new and specifically uses the parameters of the problem without any other assumptions or imposed conditions. We also provide further theoretical results, including a rate of convergence, which is also a new contribution.

We now establish additional theoretical results.

Recall that at step $t+1$

$$\lambda^{t+1} \to \max_{\lambda} \zeta_l(\lambda, \mu^t)$$

$$\mu^{t+1} \to \max_{\mu} \zeta_l(\lambda^{t+1}, \mu^t), \quad l = 1, 2, 3.$$

Let $\lambda^*_{t+1}, \mu^*_t \in$ Arg max $\zeta_l(\lambda, \mu)$ and are chosen to be the closest points to the iterates λ^{t+1}, μ^t in Arg max $\zeta_l(\lambda, \mu)$, $l = 1, 2, 3$. In order to prevent the notation from becoming cumbersome, we do omit the subscript l of the λ's and μ's. We assume for the time being that there exists the bounded set Ω_l which contains both sequences: $\{\lambda^{t+1}, \mu^t\}$ and $\{\lambda^*_{t+1}, \mu^*_t\}$. It clearly exists for $l = 1$, since $\zeta_1(\lambda, \mu)$ is a strictly concave function as can be seen from its explicit form (24). For $l = 2, 3$

we later provide a modification of the algorithm that assures that the iterates λ^{t+1}, μ^t always lie in the bounded set and, therefore, their projections on Arg max ζ_l also belong to a bounded set.

We have the following inequality for $\tau \in [0,1]$.

$$\zeta_l(\lambda^{t+1},\mu^{t+1}) \geq \zeta_l(\lambda^{t+1},\mu^t+\tau(\mu_t^*-\mu^t)) =$$

$$= \zeta_l(\lambda^{t+1},\mu^t) + \tau\nabla_{\lambda,\mu}\zeta_l(\lambda^{t+1},\mu^t) \times \left((\lambda_{t+1}^*-\lambda^{t+1})(\mu_t^*-\mu^t)\right), \quad (65)$$

$$-\frac{\bar{M}_l}{2}\tau^2(\|\lambda_{t+1}^*-\lambda^{t+1}\|^2+\|\mu_t^*-\mu^t\|^2) \geq \zeta_l(\lambda^{t+1},\mu^t)+$$

$$+\tau(\zeta_l(\lambda_{t+1}^*,\mu^*)-\zeta_l(\lambda^{t+1},\mu^t)) - \frac{\bar{M}_l}{2}\tau^2\left[\|\lambda_{t+1}^*-\lambda^{t+1}\|^2+\|\mu_t^*-\mu^t\|^2\right]$$

where $\bar{M}_l$ is a bound of the norm of the Hessian of ζ_l.

Let us prove now that for every $(\lambda^*,\mu^*) \in \Omega_l$ and for every direction $\psi = (\psi_\lambda,\psi_\mu)$ from the normal cone N_{λ^*,μ^*} to Arg max $\zeta_l(\lambda,\mu)$ at (λ^*,μ^*),

$$\zeta_l(\lambda^*,\mu^*) - \zeta_l(\lambda^*+\psi_\lambda,\mu^*+\psi_\mu) \geq \frac{A}{2}\|\psi\|^2 \quad (66)$$

where A is a positive constant and a normal cone N_{λ^*,μ^*} is defined as a set of directions ψ which satisfy

$$\psi_\lambda\cdot(\Lambda^*-\lambda^*)+\psi_\mu\cdot(M^*-\mu^*) \leq 0$$

for every $(\Lambda^*,M^*) \in$ Arg max $\zeta_l(\lambda,\mu)$.

From our definition of λ_{t+1}^*, μ_t^* it is clear that

$$\psi = \left(\lambda^{t+1}-\lambda_{t+1}^*\mu^t-\mu_t^*\right) \in N_{\lambda_{t+1}^*,\mu_t^*}.$$

Since, for every $l = 1,2,3$ the function

$$Z_l(\tau) = \zeta_l(\lambda^*+\tau\psi_\tau,\mu^*+\tau\psi_\mu), \quad \tau > 0 \quad (67)$$

is a piecewise quadratic concave function of τ (as can be seen from the explicit expressions for ζ_l). Hence, for every $\tau > 0$,

$$Z_l(\tau) \leq Z_l(0) + \frac{\partial Z_l}{\partial\tau_+}(0) + \frac{1}{2}\frac{\partial^2 Z_l}{\partial\tau_+^2}(0)\tau^2,$$

and for $\tau \in [0,\epsilon]$, $\epsilon > 0$,

$$Z_l(\tau) = Z_l(0) + \frac{\partial Z_l}{\partial\tau_+}(0) + \frac{1}{2}\frac{\partial^2 Z_l}{\partial\tau_+^2}(0)\tau^2 \quad (68)$$

where $\frac{\partial}{\partial \tau_+}$ denotes a directional derivative.

Since $(\lambda^*, \mu^*) \in \text{Arg max } \zeta_l(\lambda, \mu)$, $\frac{\partial Z_l}{\partial \tau}(0) = 0$.

Moreover, there exists a positive number $a_{\lambda^*,\mu^*,\psi} > 0$, such that

$$\frac{\partial^2 Z_l}{\partial \tau_+^2}(0) \leq -a_{\lambda^*,\mu^*,\psi}\|\psi\|^2 < 0 \tag{69}$$

since if $\frac{\partial^2 Z_l}{\partial \tau_+}(0) = 0$, then for $\tau \in [0, \epsilon]$, $Z_l(\tau) = Z_l(0) = \zeta_l(\lambda^*, \mu^*)$, which means that $(\lambda^* \in \tau\psi_\lambda, \mu^* + \tau\psi_\mu) \in \text{Arg max } \zeta_l$. This contradicts the fact that the direction (ψ_λ, ψ_μ) belongs to the normal cone.

We now introduce $\bar{a}_{\lambda^*,\mu^*} = \min_{\psi \in N_{\lambda^*,\mu^*}} a_{\lambda^*,\mu^*,\psi}$. By (67) and compactness of the normal cone, we have that $\bar{a}_{\lambda^*,\mu^*} > 0$.

Finally, we define

$$A = \min_t \{\bar{a}_{\lambda^*_{t+1},\mu^*_t}\}.$$

We now prove that $A > 0$. Indeed, assume that there exists a sequence $\{\bar{a}_{\lambda^*_{\nu+1},\mu^*_\nu}\}$ such that $\bar{a}_{\lambda^*_{\nu+1},\mu^*_\nu} \to 0$. Then by the definition of $\bar{a}_{\lambda^*\mu^*}$ and by the boundedness of Ω_l there exists a limit point (Λ^*, M^*) of this subsequence and a vector $\Psi \neq 0$ with the following properties:

$(\Lambda^*, M^*) \in \Omega_l$, $\Psi \in$ normal cone to Arg max ζ_l and by our assumption $a_{\Lambda^*,M^*,\Psi} = 0$. But, as we have shown above, the last equality is in contradiction with Ψ being a vector in a normal cone. Thus, $A > 0$. We then obtain (66) by choosing in (67) $\psi_\lambda = (\lambda^{t+1} - \lambda^*_{t+1})$ and $\psi_\mu = (\mu^{t+1} - \mu^*_{t+1})$ and then using the inequality (69) for $\tau = 1$.

We obtain then that

$$\zeta_l(\lambda^{t+1}, \mu^{t+1}) \geq \zeta_l(\lambda^{t+1}, \mu^t) + \tau(\zeta_l(\lambda^*_{t+1}, \mu^*_t) - \zeta_l(\lambda^{t+1}, \mu^t))$$

$$-\tau^2 \frac{\bar{M}_l}{A}(\zeta_l(\lambda^*_{t+1}, \mu^*_t) - \zeta_l(\lambda^{t+1}, \mu^t)). \tag{70}$$

It follows then that

$$\zeta_l(\lambda^{t+2}, \mu^{t+1}) \geq \zeta_l(\lambda^{t+1}, \mu^{t+1}) \geq \ldots \tag{71}$$

Let now

$$\delta^t = \zeta_l(\lambda^*_{t+1}, \mu^*_{t+1}) - \zeta_l(\lambda^{t+1}, \mu). \tag{72}$$

Using then (70), (71), and (72) we obtain

$$\zeta_l(\lambda^*_{t+2}, \mu^*_{t+1}) - \zeta_l(\lambda^{t+2}, \mu^{t+1}) \leq \zeta_l(\lambda^*, \mu^*) - \zeta_l(\lambda^{t+1}, \mu^t) -$$

$$-\tau(\zeta_l(\lambda^*, \mu^*) - \zeta_l(\lambda^{t+1}, \mu^t)) + \tau^2 \frac{\bar{M}_l}{A}(\zeta_l(\lambda^*, \mu^*) - \zeta_l(\lambda^{t+1}, \mu^t)) \tag{73}$$

or

$$\delta^{t+1} \leq \delta^t(1 - \tau + \tau^2 \frac{\bar{M}_l}{A}) \tag{74}$$

for $\tau \in [0, 1]$.

Minimizing (74) with respect to τ, we obtain:

$$\tau_{\min} = \frac{A}{2\bar{M}_l}. \tag{75}$$

Hence,

$$\delta^{t+1} \leq \delta^t(1 - \frac{A}{4\bar{M}_l}) \tag{76}$$

where $1 - \frac{A}{4\bar{M}_l} < 1$.

Therefore, if for some fixed $\bar{\epsilon} > 0$, we use the stopping criterion $\delta^{\bar{T}} \leq \bar{\epsilon}$, then the number of steps $\bar{T}$ for convergence, is given by:

$$\bar{T} = \frac{\ln\left[\frac{\bar{\epsilon}}{\delta^0}\right]}{\ln\left[(1 - \frac{A}{4\bar{M}_l})\right]}. \tag{77}$$

Observe that the # of iterations $\bar{T}$ is additive with respect to $\bar{\epsilon}$. Hence if we decrease $\bar{\epsilon}$ by a factor of 10, we should expect to see only an additive increase in the number of iterations.

For each iteration, cf. Eydeland and Nagurney (1989), (assuming that $m = n$) each demand/supply exact equilibration algorithm takes

$$7n + n \ln n + 2n \quad \text{operations}.$$

Hence, it takes

$$n(9n + n \ln n) \quad \text{operations}$$

for all n rows/columns to be equilibrated. The overall number of operations N (and correspondingly the overall CPU time) is then proportional to

$$N = \bar{T}(n^2)(9 + n \ln n).$$

Note, that if there are p processors available, where we assume that $p \leq n$ then

$$N_p = \frac{\bar{T}(n^2)(9 + n \ln n)}{p}.$$

In particular, for $p = n$, we have that

$$N_n = \bar{T}n(9 + n \ln n).$$

We shall now prove how to ensure that (λ^{t+1}, μ^t) belongs to a bounded set in the case of $l = 2, 3$.

It is clear that there exists a d_{max} such that if $\lambda_i + \mu_{j'}$ is larger than d_{max}, then ζ_l is $< \zeta_l^0$, which cannot be true since at any step $\zeta_l \geq \zeta_l^0$ because we are maximizing ζ_l. Also, it is clear that if $\lambda_i + \mu_{j'} < -d_{max}$, then $x_{ij'} = 0$.

If at step t, then, we have that $x_{ij'}^t > 0$, then $\lambda_i + \mu_{j'} > -d_{max}$. Hence, we can conclude that

$$-d_{max} < \lambda_i + \mu_{j'} < d_{max}. \tag{78}$$

Note that d_{max} depends only on the given data of the problem and not on t.

Each iterate $(x^t, \lambda^{t+1}, \mu^t)$ of our procedure defines a graph G^t whose nodes (ij') are connected only if $x_{ij'} \neq 0$. In this graph we can introduce the definition of adjacency of two edges: the edge (ij') is adjacent to (kl') if either $i = k$ or $l' = j'$. Thus, we have a definition of a new graph G^{t^*} whose nodes, corresponding to edges in G^t, are connected if and only if the edges in G^t have a common endpoint. Having a definition of connectedness of two nodes in G^{t^*} (edges in G^t) we may now define the connected component in G^{t^*} in a standard way. It is clear from the definition of $\zeta_l (l = 2, 3)$ that within a connected component one can add a certain constant to λ_i's and subtract the same constant from $\mu_{j'}$ without changing the values of ζ_l. Moreover, by (78), if edges (ij') and (kl') belong to a connected component then

$$|\lambda_i - \lambda_k| < 2nd_{max}$$

$$|\mu_{j'} - \mu_{l'}| < 2nd_{max}.$$

Combining these properties together we define the following modification of our algorithm which would keep the iterates (λ^{t+1}, μ^t) in the bounded set.

Modified Algorithm:

Choose a large $R > 0$.

Let λ^t, μ^t be known.

If all λ_i's are $< R$, continue to the next t-th step.

If there exists a $\tilde{\lambda}_i$ such that $|\tilde{\lambda}_i| > R$, then subtract $\tilde{\lambda}_i$ from all λ_i's in the connected component and add it to all $\mu_{j'}$'s in this connected component. This should bring all of the λ's and μ's in the cube $[-2nd_{max}, 2nd_{max}]$. Then check other connected components.

By the properties discussed above this modification does not change the values of ζ_l. Moreover, since $\tilde{\lambda}_i^{modified} = 0$ all other λ_i's in the connected component will be less than $2nd_{max}$ in absolute values. Clearly, $|\mu_{j'}^{modified}|$ are bounded in the same way.

We now present the splitting equilibration algorithm for the general problem.

3.2. The Splitting Equilibration Algorithm for General Problems

SEA for general problems solves a series of diagonal problems as outlined in Section 3.1. The diagonal problems, in turn, are constructed via the projection method of Dafermos (1982, 1983) which is based on variational inequality theory. For a brief introduction to variational inequality theory and associated applications, see Nagurney (1987). In particular, the projection method constructs a series of quadratic programming problems which are simpler than the original problem. It uses fixed matrices and modifies only the fixed linear terms in the corresponding objective functions. In particular, we select as the fixed matrices the diagonals of general matrices A, B, and G. Hence, only the linear terms are updated from iteration to iteration.

We now present the splitting equilibration algorithm for the general constrained matrix problem, (1) through (4). For a graphical depiction of SEA for general problems, we refer the reader to Figure 4.

3.2.1. SEA for General Constrained Matrix Problems with Unknown Row and Column Totals

Step 0: Initialization Step

Start with any feasible (s, x, d), i.e., one which satisfies constraints (2), (3), and (4). Set $t = 1$ and let $0 < \alpha \leq 1$.

Step 1: Projection Step

Given $(s^{t-1}, x^{t-1}, d^{t-1})$, find (s^t, x^t, d^t) by solving the following problem:

$$\text{Minimize } \frac{1}{2}s^T\tilde{A}s + \alpha(-\tilde{A}s^{t-1} - As^0 + As^{t-1})^T s$$

$$+\frac{1}{2}x^T\tilde{G}x + \alpha(-\tilde{G}x^{t-1} - Gx^0 + Gx^{t-1})^T x$$

$$+\frac{1}{2}d^T\tilde{B}d + \alpha(-\tilde{B}d^{t-1} - Bd^0 + Bd^{t-1})^T d \qquad (79)$$

subject to constraints (2), (3), and (4), via SEA for Diagonal Problems (see Section 3.1), where $\tilde{A}, \tilde{G}$, and $\tilde{B}$ denote the diagonal matrices diag(A), diag(G), and diag(B), respectively.

Step 2: Convergence Verification

If $|x_{ij}^t - x_{ij}^{t-1}| \leq \epsilon$, for all i, j, then stop; otherwise, set $t = t + 1$, and go to Step 1.

The general splitting equilibration algorithm applied to both the SAM problem and the problem with fixed row and column totals can be constructed in an analogous manner.

TABLE 1
Computational experience with SEA on large-scale diagonal quadratic constrained matrix problems*

$m \times n$	Number of non-zero x_{ij}^0 variables	CPU time (seconds)
750 × 750	562,500	204.7476
1000 × 1000	1×10^6	483.2065
2000 × 2000	4×10^6	3,823.2139
3000 × 3000	9×10^6	13,561.5703

* CPU time based on a single example.

TABLE 2
Computational experience with SEA on United States input/output matrix datasets

Dataset	CPU time (seconds)
IOC72a	18.6697
IOC72b	18.9923
IOC72c	25.6035
IOC77a	13.6168
IOC77b	19.1338
IOC77c	30.2037
IO72a	333.2691
IO72b	438.3519
IO72c	335.6124

TABLE 3
Computational experience with SEA on social accounting matrix datasets

Dataset	Number of accounts	Number of transactions	CPU time (seconds)
STONE	5	12	.0024
TURK	8	19	.0210
SRI	6	20	.009
USDA82E	133	17,689	5.7598
S500	500	250,000	28.99
S750	750	562,500	52.60
S1000	1000	1,000,000	95.08

TABLE 4
Computational experience with SEA on United States migration tables

Dataset	CPU time (seconds)
MIG5560a	1.5935
MIG5560b	4.1367
MIG5560c	.8932
MIG6570a	1.2915
MIG6570b	3.9714
MIG6570c	.8203
MIG7580a	3.5168
MIG7580b	9.1067
MIG7580c	.8041

TABLE 5
Computational experience with SEA on spatial price equilibrium problems

$m \times n$	Number of variables	CPU time (seconds)
SP50 × 50	250	1.3822
SP100 × 100	10,000	11.2621
SP250 × 250	62,500	129.4597
SP500 × 500	250,000	540.7056
SP750 × 750	562,500	1589.0613

* CPU time based on a single example.

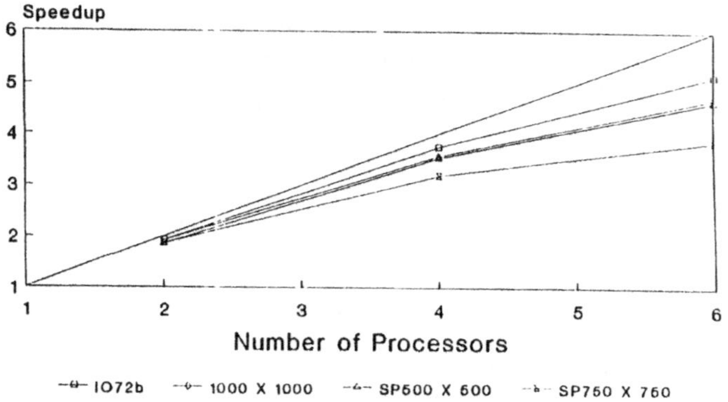

Fig. 5. Speedup for SEA on diagonal problems.

TABLE 6
Parallel speedup and efficiency measurements for SEA on diagnonal problems

Example	N	S_N	E_N
IO72b	2	1.93	96.5%
	4	3.74	93.5%
	6	5.15	85.8%
1000 × 1000	2	1.93	96.5%
	4	3.57	89.4%
	6	4.71	78.5%
SP500 × 500	2	1.86	92.85%
	4	3.52	88.10%
	6	4.66	77.75%
SP750 × 750	2	1.87	93.79%
	4	3.19	79.80%
	6	3.86	64.34%

4. Computation of Large-Scale Diagonal Constrained Matrix Problems

In this and the subsequent Sections we describe the computational experiments conducted and the results obtained for the splitting equilibration algorithm (SEA) on large-scale quadratic constrained matrix problems. We begin with computational experience on diagonal problems and then turn to the solution of general problems. For each class of problems we first present the results of serial computations and then those of parallel computations. All of the computational experiments were conducted on the IBM 3090-600E at the Cornell National Supercomputer Facility (CNSF).

4.1. Serial Experiments

In this Section we investigated the computational efficiency of the SEA algorithm on the class of diagonal constrained matrix problems outlined in Section 2. All of the programs used throughout the study were coded in FORTRAN and run on the IBM 3090-600E at the CNSF. The serial programs were compiled under VS FORTRAN at optimization level (3) running under VM/XA 5.5. The CPU times are exclusive of input and output, but include initialization times.

TABLE 7
Computational comparisons of SEA, RC, and B-K on general quadratic constrained matrix problems with 100% dense G matrix

Dimension of G	Number of of runs	CPU time (seconds) SEA	RC	B-K
100 × 100	10	.0194	.1270	.7725
400 × 400	10	.5694	1.8373	78.9557
900 × 900	2	2.9767	9.5129	1458.3820
2500 × 2500	1	21.4607	71.4807	–
4900 × 4900	1	81.2640	428.8780	–
10000 × 10000	1	353.6885	1305.5940	–
14400 × 14400	1	1254.731	3000.5200	–

TABLE 8
Computational experience with SEA on general constrained matrix problems consisting of United States migration tables with 100% dense G matrices – dimension of G – 2304 × 2304

Dataset	CPU time (seconds)
GMIG5560a	23.16
GMIG5560b	22.99
GMIG6570a	23.57
GMIG6570b	23.28
GMIG7580a	28.73
GMIG7580b	23.49

TABLE 9
Parallel speedup and efficiency measurements for SEA and RC on general problems

Example	N	S_N	E_N
SEA	2	1.82	90.77%
10000 × 10000	4	2.62	65.49%
RC	2	1.75	87.7%
10000 × 10000	4	2.24	55.9%

4.1.1. Computational Experience with SEA

In this Section we studied the performance of SEA on very large problems with fixed row and column totals. The examples were generated as follows. We generated matrix examples ranging in size from 750 rows $\times$750 columns through 3000 rows $\times$3000 columns with the percentage of positive $x^0_{ij'}$'s at 100%. Each non-zero $x^0_{ij'}$ was generated uniformly in the range $[.1, 10000]$, to simulate the wide spread of the initial data which are characteristic of both input/output and social accounting matrices. The weighting terms, the $\gamma_{ij'}$'s were set to $\frac{1}{x^0_{ij'}}$. In the examples, we set each row total $s^0_i = 2\sum_{j'} x^0_{ij'}$, and each column total $d^0_{j'} = 2\sum_i x^0_{ij'}$.

SEA was implemented in accordance with the suggestions and theoretical guidance for equilibration algorithms contained in Eydeland and Nagurney (1989). In particular, each row equilibrium subproblem and each column equilibrium subproblem was solved via exact equilibration. Since exact equilibration requires sorting and since the arrays to be sorted are, typically, in the applications considered here substantially larger than one hundred elements, the sorting procedure used in the implementation of SEA for fixed row and column totals was HEAPSORT. The ϵ in the convergence test was set at .01. The results of the numerical experiments are reported in Table 1.

The examples solved ranged from 562,500 to 9 million non-zero initial matrix elements. As can be seen from Table 1, the smallest example with 562,500 variables required only minutes of CPU time on a serial machine, whereas the largest, with 9,000,000 variables required approximately $3\frac{1}{2}$ hours of CPU time. On the basis of these runs we now consider applications with fewer than 1 million variables to be solvable feasibly in a reasonable time-frame in a serial manner on a machine such as the IBM 3090-600E when an efficient algorithm such as SEA is used for the computation.

4.1.2. Experiments on Input/Output Matrices, Social Accounting Matrices, Migration Tables, and Spatial Price Equilibrium Problems

In this Section we provide further numerical results with SEA on real-world economic and demographic datasets. The datasets include input/output matrices, social accounting matrices, migration tables, and spatial price equilibrium problems.

We now briefly describe the datasets. The computational results are reported in Tables 2, 3, 4, and 5. In Table 2 we report the results of the performance of SEA on input/output matrices with known row and column totals. In Table 3 we report the results of SEA on social accounting matrices, in which the row and column totals must balance and must be estimated, as well. In Table 4 we report the results of SEA on migration tables, in which the row and column totals are also to be estimated. In Table 5 we report the results of SEA's performance on spatial price equilibrium problems, which, as discussed in Section 2, are equivalent to constrained matrix problems in which both row and column totals need to be estimated.

The first set of three examples was constructed from an aggregated 1972 input/output matrix of construction activity in the United States consisting of 205 rows ×205 columns. This I/O matrix retained the construction sectors in the United States in detail, and aggregated those sectors in the United States in which the construction inputs were zero or negligible. The first example, IOC72a, was formed by generating a 10% growth factor, while the second, IOC72b, by generating a 100% growth factor. The percentage of non-zero $x^0_{ij'}$'s was 52%. The third datapoint, termed, IOC72c, consisted of the average of 10 examples, where each example consisted of the 1972 matrix perturbed by a randomly generated additive term in the range $[1, 10]$.

The second series of three examples was constructed from an aggregated 1977 input/output matrix of construction activity in the United States consisting of 205 rows ×205 columns in the same manner as those in the first series, and are called, respectively, IOC77a, IOC77b, and IOC77c. Examples IOC77a and IOC77b consisted of 58% non-zero elements in the X^0 matrix.

The third and final series of I/O examples was constructed from a 1972 input/output matrix for the United States consisting of 485 rows ×485 columns in a manner similar to the examples in the first two series. These examples are referred to, respectively, as IO72a, IO72b, and IO72c. These matrices were the sparsest, with only 16% non-zero elements in the X^0 matrix.

As can be seen from Table 2, all of the examples, with the exception of the largest set based on the disaggregated 1972 I/O matrix, required only seconds of CPU time for computation of the solution via SEA. The largest examples required less than 8 minutes.

We now describe the SAM estimation problems. The SAM estimation problems selected were of various sizes. The first four examples were selected because they represented real economic datasets. The last three examples were generated to introduce large-scale SAM problems of a size larger than heretofore considered computationally tractable.

The smallest example, STONE, had also been solved in Byron (1978). The USDA82E example, was a perturbed SAM developed at the United States Department of Agriculture for 1982 (For a description of its development, we refer the reader to Hanson and Robinson (1989)). It was perturbed in order to make it fully dense, and a 'difficult' problem. The example SRI is a perturbed example of the SAM for Sri Lanka for 1970 contained in King (1985). TURK is a perturbed SAM for the 1973 Turkish economy discussed in Dervis, De Melo, and Robinson (1982). Examples S500, S750, and S1000 are large-scale randomly generated SAM's. The convergence tolerance was set at $\epsilon = .001$.

As can be seen from Table 3, SEA was very efficient, computing the solution for the first five examples in only fractions of a CPU second. The largest economic dataset, USDA82E, required only several CPU seconds for convergence of SEA. The largest three problems demonstrate the scale of SAM estimation problems that are now solvable in a reasonable time-frame, even in a serial environment, provided that a robust and efficient algorithm such as SEA is utilized.

In Table 4 we report the performance of SEA on United States migration tables for different time periods. The objective function used was again diagonal, of the form (5). The rows of each migration table represent the origin states, and the columns, the destination states. Alaska, Hawaii, and Washington, DC were removed, creating tables with 48 rows and 48 columns.

The first set of three migration table examples in Table 4 was constructed from a 1955–1960 U.S. state to state migration table. The first example, MIG5560a, was formed by generating a distinct random growth factor for each row and column total in the range 0–10%, and by then using the resultant as the s_i^0, or $d_{j'}^0$ (cf. (5)), respectively. All of the weights were set equal to one. The second example, MIG5560b, was formed by generating a distinct growth factor again for each original row and column total, but now within the larger range of 0–100%. The third example, MIG5560c was formed by keeping the s_i^0's and $d_{j'}^0$'s equal to the sums of the corresponding matrix entries in the original table; the X^0 matrix was then constructed by perturbing each element randomly by 0–10%.

The second set of three examples was constructed from the 1965–1970 U.S. state to state migration table in a manner similar to the first set, and these examples were named: MIG6570a, MIG6570b, and MIG6570c, respectively.

The third and final set of migration tables was constructed from a 1975–1980 U.S. state to state migration table in a manner similar to the examples in the first two sets, and were named MIG7580a, MIG7580b, and MIG7580c, respectively.

As can be seen from Table 4, the migration table examples were computed in only seconds of CPU time. The examples with the greater growth factor were more difficult to solve (as expected) than the corresponding ones with the smaller growth factor range. The examples with the perturbed matrix entries were solved most quickly.

Finally, we turned to the computation of classical spatial price equilibrium problems. Spatial price equilibrium models have been widely applied to the study of agricultural and energy markets. Specifically, we consider spatial price equilibrium problems, characterized by linear supply price, demand price, and transportation cost functions which are also separable. We generated large-scale spatial price equilibrium problems ranging in size from 50 supply markets and 50 demand markets, with a total of 250 variables, to 750 supply markets and 750 demand markets, with a total of 562,500 variables. The convergence tolerance was $\epsilon = .01$. As can be seen from Table 5, SEA converged for all of the examples. Although serial equilibration algorithms have been proposed for such problems by Dafermos and Nagurney (1989) and Eydeland and Nagurney (1989), the problem with 750 supply markets and 750 demand markets represents the largest of this class solved to date.

4.2. Parallel Experiments

In this Section we describe our experiences concerning a parallel implementation of SEA for the computation of large-scale diagonal quadratic constrained matrix problems. The experiments were carried out on the IBM 3090-600E, a shared memory machine, using its full multiprocessor capabilities.

The SEA algorithm (diagonal version) was embedded with the parallel constructs provided by Parallel FORTRAN (PF) for purposes of task allocation (cf. Figure 2). Task allocation was required for the row equilibration phase and the column equilibration phase, with cycling between the two phases until the convergence criterion was satisfied.

For the computational testing, SEA was compiled using the Parallel FORTRAN (PF) compiler, optimization level (3). We selected four previously solved examples for the parallel tests, specifically, IO72b from Table 2, the 1000×1000 example from Table 1, and SP500 $\times$ 500 and SP750 $\times$ 750 from Table 5. Recall that both IO72b and 1000×1000 assume fixed row and column totals, whereas SP500 $\times$ 500 and SP750 $\times$ 750 are spatial price equilibrium problems, isomorphic to constrained matrix problems with unknown row and column totals.

The speedup measure for N processors was defined as follows:

$$\text{Speedup} \quad S_N = \frac{T_1}{T_N},$$

where T_1 is the elapsed time to solve the problem using the serial implementation of SEA on a single processor, and T_N is the elapsed time to solve the problem using the parallel implementation of SEA on N processors.

The efficiency measure for N processors was defined as:

$$\text{Efficiency} \quad E_N = \frac{T_1}{T_N \times N}.$$

In Table 6 we report the speedup measurements and the corresponding efficiencies obtained. These measurements were obtained in a standalone environment. The speedup measurements are then displayed graphically in Figure 5.

SEA required 2 iterations for convergence for example IO72b and only 1 iteration for the 1000×1000 example. As can be seen from Table 6, SEA exhibited identical speedups of 1.93, or, equivalently, efficiencies of 96.5% on both diagonal examples when 2 CPU's were used. In the case of 4 CPU's SEA applied to IO72b exhibited 93.5% efficiency, whereas the 1000×1000 example induced an efficiency of 89.4%. For 6 CPU's, SEA again exhibited the higher speedup for IO72b of 5.15, at an efficiency of 85.5%. This difference in relative speedups can be explained by the portion of time spent in the serial phase which consists solely of the convergence criterion verification stage for diagonal SEA. The larger example, 1000×1000 required more serial time spent in this serial phase. Although SEA required only 1 iteration for this example, the serial convergence step is on the order of m^2 operations, where m in this example is equal to 1000. On the other hand, even though SEA required 2 iterations in the case of IO72b, since m in this example is equal to 485, the total time spent in the serial phase would be approximately 50% less in the smaller example. Enhanced speedups may be obtained by verifying convergence not after every iteration, as was done in these tests, but after every other iteration when the number of iterations is small or by implementing the convergence phase in parallel. Nevertheless, practitioners are interested in the solutions themselves, and, therefore, convergence verification is a vital step.

SEA required 84 iterations for convergence of SP500 $\times$ 500 and 104 iterations for convergence of SP750 $\times$ 750, where the convergence check was done after every other iteration. In these 'elastic' examples, convergence verification again comprised the only serial phase, and was of the order m^2. Again, the larger example required greater time in the serial phase of convergence verification. Here, enhanced speedups may be obtained by verifying convergence, say, after every five iterations and/or by implementing the convergence step in parallel. The explanation of the greater number of iterations required for convergence of SEA in the case of the elastic examples versus the fixed examples may lie in the initialization phase; $\mu = 0$ may be closer to the optimal for the latter examples, than for the former ones.

5. Computation of Large-Scale General Constrained Matrix Problems

In this Section we describe the computation of large-scale general quadratic constrained matrix problems, formulated in Section 2. Recall that the general quadratic constrained matrix problem is computed via the iterative solution of diagonal constrained matrix problems. SEA for the general case needs substantial storage since the G matrix may, in fact, be fully dense. Hence, for an initial matrix X^0, consisting of 100 rows $\times$100 columns, the corresponding G matrix would be of dimension $10,000 \times 10,000$.

5.1. Serial Experiments

In this Section we provide the results of serial experimentation. The SEA, RC (Nagurney, Kim, and Robinson (1990)), and B-K (Bachem and Korte (1978)) programs for the general quadratic problems were coded in FORTRAN, compiled under VS FORTRAN at optimization level (3) running under VM/XA 5.5. Details of the implementation of the B-K algorithm are given in Nagurney, Kim, and Robinson (1990).

5.1.1. Computational Comparisons of SEA, RC, and B-K

Similar to the general SEA, the RC algorithm is also an equilibration algorithm based on the projection method, and involves the iterative solution of diagonal

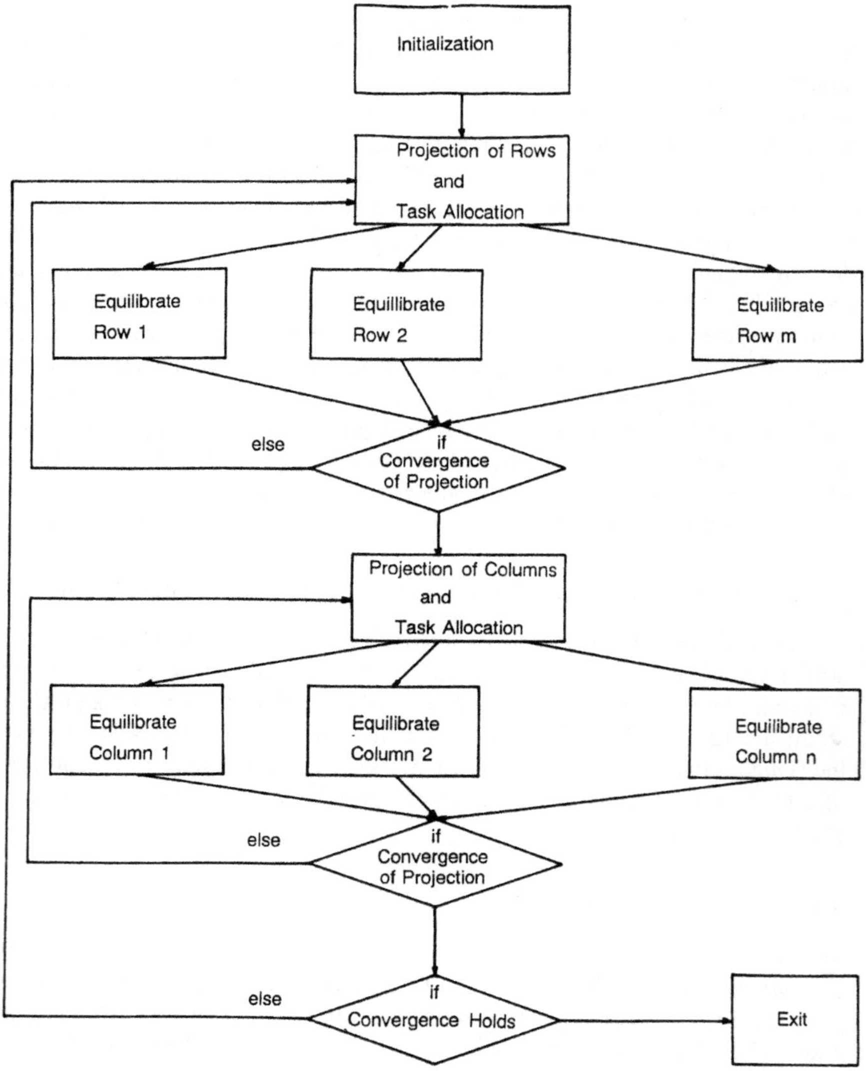

Fig. 6. Flowchart of the RC algorithm – the general case.

constrained matrix problems. It, however, first considers the general objective function (1) subject to only the row constraints, and then subject to the column constraints. A graphical representation of the RC algorithm for the general problem is presented in Figure 6.

Our computational comparisons of SEA versus RC and B-K are conducted for the general constrained matrix problem with fixed row and column totals, since both RC and B-K were designed for this class of constrained matrix problems.

The matrix G was generated to be symmetric and strictly diagonally dominant, which ensured positive definiteness, with each diagonal term generated in the range $[500, 800]$, but allowing for negative off-diagonal elements to simulate variance-covariance matrices. Each element of the linear term coefficients in the expansion of (1) was generated uniformly in the range $[100, 1000]$. The same convergence criterion was used for B-K, RC, and SEA, with $\epsilon' = .001$.

The implementation of both SEA and general RC was done in accordance with the guidelines for the implementation of equilibration algorithms contained in Eydeland and Nagurney (1989) . The general problems computed with both SEA and RC ranged in size of X^0 matrices from 10×10 to 120×120, with the corresponding G matrices ranging in size from 100×100 to 14400×14400, respectively. The STRAIGHT INSERTION SORT was used for the implementation of exact equilibration, since the arrays to be sorted ranged in length from 10 elements to 120 elements. Table 7 presents computational comparisons of SEA versus RC and B-K on general quadratic constrained matrix problems solved in Nagurney, Kim, and Robinson (1990) with 100% dense G matrices. As can be seen from Table 7, SEA outperformed RC by a factor of 3 to 4, and outperformed B-K by as much as two orders of magnitude. The larger problems were not solved using B-K because it became prohibitively expensive to do so.

5.1.2. Computational Experience with SEA on Migration Tables

In this Section we considered United States migration tables for different time periods, for which the constrained matrix formulation with objective function (1) was again used. The weighting matrix G was generated in the same manner as in Section 4.1.1. These United States migration tables, from which we constructed the examples, consisted of 48 rows and 48 columns. The rows of each migration table represented the origin states and the columns the destination states. Alaska, Hawaii, and Washington, DC, were removed, thus creating tables with 48 rows and columns. The G matrices were, hence, of dimension 2304×2304. The examples, reported in Table 8, were as follows.

The first set of two examples, GMIG5560a and GMIG5560b, were based on the 1955–1960 U.S. state to state migration table. GMIG5560a consisted of the baseline table with row and column totals being fixed and consisting of a growth factor in the range 0–10%. GMIG5560b, then, in addition, to the row and column total perturbations, had each individual matrix entry perturbed by a distinct growth factor, also in the range 0–10%.

The second and third sets of two examples each were based on the 1965–1970 and the 1975–1980 U.S. state to state migration tables, respectively, and were constructed in a manner similar to the examples in the first set. As can be seen

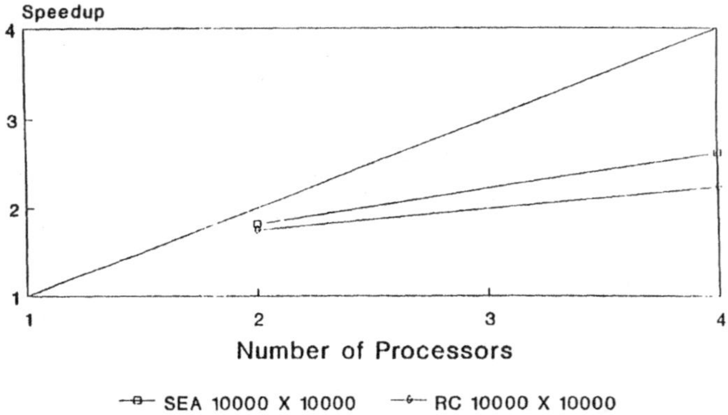

Fig. 7. Speedup for SEA vs RC on general problems.

from Table 8 below, all of the examples were solved via SEA in approximately 25 seconds of CPU time with ϵ' set to .001.

5.2. Parallel Experiments with SEA

For purposes of parallel experimentation, we selected the 10000×10000 example contained in Table 7, which had also been solved using a parallel implementation of RC in Nagurney, Kim, and Robinson (1990). The parallel implementation of the SEA algorithm (cf. Figure 4) used Parallel FORTRAN (PF) as did the RC algorithm. For the computational testing, both SEA and RC were compiled using the PF compiler, optimization level 3.

Recall that SEA resolves general quadratic constrained matrix problems into series of diagonal row and column equilibrium subproblems. In RC, the parallel phases of exact row equilibration and exact column equilibration are separated via the serial phase of projection method convergence verification, which adds a serial phase not encountered in the parallelization of the SEA algorithm (cf. Figures 4 and 6), in which convergence verification of the projection method is only done once.

The example was solved in 2 iterations of general RC and in 1 iteration via SEA. The 10000×10000 example required for RC in the first iteration, 4 iterations of the projection method for row equilibration and 3 iterations of the projection method for column equilibration, whereas in the second iteration, 4 iterations for both equilibrations were required. SEA, on the other hand, besides requiring only a single outer iteration, only required two inner iterations.

Table 9 contains the speedups and the efficiencies obtained, whereas Figure 7 graphically depicts the speedups for both SEA and RC. These speedups and efficiencies were obtained in a standalone environment.

As can be seen from Table 9, SEA exhibited higher speedups than RC for the example. In the case of 2 CPU's, SEA exhibited a speedup of 1.82 versus

1.75 obtained with RC; in the case of 4 CPU's, SEA exhibited a speedup of 2.62 versus 2.24 obtained with RC. Hence, SEA registered an improvement in absolute efficiency of 3.03% in the case of 2 CPU's and 9.59%, in the case of 4 CPU's. These parallel results, and the serial results contained in Section 5.1.1, strongly suggest that the new Splitting Equilibration Algorithm is better suited for parallelization than RC and more effective for the computation of large-scale constrained matrix problems, in either a serial or a parallel environment.

Acknowledgements

The first author's work was supported by NSF Grant RII-880361 under the NSF VPW program and by a 1990 Faculty Fellowship Award from the University of Massachusetts at Amherst while she was a visiting faculty member at the Transportation Systems Division, the Operations Research Center, and the Sloan School of Management at the Massachusetts Institute of Technology. The cordiality and hospitality of the host institution are warmly appreciated. The second author's work was supported by NSF Grant: DMS-8602316 and completed while he was on sabbatical leave at the Courant Institute, New York University.

The research was conducted at the Cornell National Supercomputer Facility, a resource of the Center for Theory and Simulation in Science and Engineering at Cornell University, which is funded in part by the National Science Foundation, New York State, and the IBM Corporation.

The authors would like to thank Karen Polenske and Nic Rockler of the Department of Urban Planning at MIT for providing the input/output datasets, Waldo Tobler of the Department of Geography at the University of California at Santa Barbara for providing the migration tables, and Kenneth Hanson of the US Department of Agriculture, Economic Research Service, for providing the USDA social accounting matrix.

The authors would like to thank Dae-Shik Kim for assistance with the numerical runs and Alan G. Robinson of the Department of General Business and Finance in the School of Management at the University of Massachusetts at Amherst for stimulating conversations during early phases of this research.

The authors are indebted to Francesca Verdier of the Cornell Theory Center for assistance in the standalone runs.

References

Bacharach, M. (1970), *Biproportional scaling and input-output change*, Cambridge University Press, Cambridge, UK.

Bachem, A. and Korte, B., (1978), 'Algorithm for quadratic optimization over transportation polytopes,' *Zeitschrift fur Angewandte Mathematik und Mechanik*, **58**: T459–T461.

Bertsekas, D. P. and Tsitsiklis, J. N., (1989), *Parallel and Distributed Computation: Numerical Methods*, Prentice-Hall, Englewood Cliffs, New Jersey.

Byron, R. P., (1978), 'The estimation of large social accounts matrices,' *Journal of the Royal Statistical Society Series A*, **141**: 359–369.

Cottle, R. W., Duvall, S. G., and Zikan, K., (1986), 'A lagrangean relaxation algorithm for the constrained matrix problem,' *Naval Research Logistics Quarterly*, **33:** 55–76.

Dafermos, S., (1982), 'The general multimodal network equilibrium problem with elastic demand,' *Networks* **12:** 57–72.

Dafermos, S., (1983), 'An iterative scheme for variational inequalities,' *Mathematical Programming*, **26:** 40–47.

Dafermos, S. and Nagurney, A., (1989), 'Supply and demand equilibration algorithms for a class of market equilibrium problems,' *Transportation Science*, **23:** 118–124.

Deming, W. E. and Stephan, F. F., (1940),'On a least-squares adjustment of a sampled frequency table when the expected marginal totals are known,' *Annals of Mathematical Statistics*, **11:** 427–444.

Dervis, K., De Melo, J., and Robinson, S., (1982), *General equilibrium models for development policy*, Cambridge University Press, Cambridge, United Kingdom.

Enke, S., (1951), 'Equilibrium among spatially separated markets: solution by electric analogue,' *Econometrica* , **19**, 40–48.

Eydeland, A. and Nagurney, A., (1989),'Progressive equilibration algorithms: the case of linear transaction costs,' *Computer Science in Economics and Management*, **2**: 197–219.

Friedlander, D., (1961), 'A technique for estimating a contingency table given the marginal total and some supplementary data,' *Journal of the Royal Statistical Society A*, **124:** 412–420.

Hanson, K. A. and Robinson, S., (1989), 'Data, linkages, and models: U.S. national income and product accounts in the framework of a social accounting matrix,' Agriculture and Rural Economy Division, Economic Research Service, U. S. Department of Agriculture. Staff Report No. AGES 89–5.

Harrigan, F. and Buchanan, I., (1984), 'Quadratic programming approach to input-output estimation and simulation,' *Journal of Regional Science*, **24, no. 3:** 339–358.

Judge, G. G. and Yancey, T. A., (1986),*Improved methods of inference in econometrics*, North-Holland.

King, B. B., (1985), 'What is a SAM?', in *Social accounting matrices: a basis for planning*, G. Pyatt and J. I. Round (eds.), The World Bank, Washington, DC.

Klincewicz, J., (1989), 'Implementing an exact Newton method for separable convex transportation problems,' *Networks*, **19:** 95–105.

Miller, R. E. and Blair, P. D., (1985),*Input/output analysis: foundations and extensions*, Prentice-Hall, Englewood Cliffs, N.J.

Mohr, M., Crown, W. H., and Polenske, K. E., (1987), 'A linear programming approach to solving infeasible RAS problems,' *Journal of Regional Science* , **27, no. 4:** 587–603.

Mosteller, F. and Tukey, J. W., (1977),*Data analysis and regression*, Addison-Wesley Publishing Co., Inc.

Nagurney, A., (1987),'Competitive equilibrium problems, variational inequalities, and regional science, ' *Journal of Regional Science* , **27:** 503–514.

Nagurney, A., (1989),'An algorithm for the solution of a quadratic programming problem with application to constrained matrix and spatial price equilibrium problems,' *Environment and Planning A*, **21:** 99–114.

Nagurney, A., Kim, D. S., and Robinson, A. G., (1990), 'Serial and parallel equilibration of large-scale constrained matrix problems with application to the social and economic sciences,' *The International Journal of Supercomputer Applications* , **4.1**, Spring, 49–71.

Nagurney, A. and Robinson, A. G., (1989), 'Algorithms for quadratic constrained matrix problems,' to appear in *Mathematical and Computer Modelling*.

Ohuchi, A. and Kaji, I., (1984),'Lagrangean dual coordinatewise maximization for network transportation problems with quadratic costs,' *Networks*, **14:** 525–530.

Polenske, K. E., (1980),*U. S. multiregional input-output accounts and model*, Lexington Books, Lexington, MA.

Pyatt, G. and Round, J. I., (1985),(eds.), *Social accounting matrices: a basis for planning*, The World Bank, Washington, DC.

Samuelson, P. A., (1952),'A spatial price equilibrium and linear programming,' *American Economic Review* , **42:** 283–303.

Stone, R., (1951), 'Simple transaction models, information, and computing,' *The Review of Economic*

Studies , **XIX (2), 49:** 67–84.

Stone, R., (1962),'Multiple classifications if social accounting,' *Bulletin de l'Institut International de Statistique,* **39:** 215–233.

Takayama, T. and Judge, G. G., (1971),*Spatial and temporal price and allocation models*, North-Holland, Amsterdam.

Van der Ploeg, F., (1982), 'Reliability and adjustment of sequences of large economic accounting matrices,' *Journal of the Royal Statistical Society A,* **145:** 169–184.

Van der Ploeg, F., (1988), 'Balancing large systems of national accounts,' *Computer Science in Economics and Management,* **1:** 31–39.

Van der Sluis, A., (1969), 'Condition numbers and equilibration of matrices,' *Numerische Mathematik,* **14:** 14–23.

NONSTATIONARY MODEL SOLUTION TECHNIQUES AND THE USA ALGORITHM

P.G. FISHER
Bank of England, UK

and

A.J. HUGHES HALLETT
University of Strathclyde, Glasgow
and
Centre for Economic Policy Research, London, UK

Abstract. A variety of iterative techniques are available for solving large, sparse nonlinear equation systems. This paper reviews first-order, non-stationary iterations which automatically accelerate convergence. In particular we compare automatic Fast-Gauss-Seidel methods, the USA algorithm and fixed parameter Fast-Gauss-Seidel iterations for solving large-scale nonlinear macroeconomic models. Specific attention is drawn to their robustness properties and computational costs. On both grounds, the USA algorithm is found to compare unfavourably for a selection of models of the UK economy.

1. Introduction

Economic models are mostly used for making a series of forecasts or simulations over different time periods and for a range of small changes in the external conditions or base assumptions. Model solution techniques therefore need to be computationally efficient in order to reduce the cost of repeatedly solving a large sparse system. Perhaps more important, they need to be *robust* in the sense that the speed of convergence is not reduced significantly by the 'small' variations in the exogenous variables, model structure, or start values which are needed to generate a sequence of (policy) simulations or forecasts.

Research in this area has therefore concentrated on determining the convergence conditions for standard solution techniques applied to equation systems which have none of the usual features of physical systems, such as symmetry, diagonal dominance, or nonnegativity. A second theme has been the search for ways of reducing the computational costs of model solution and maximising the convergence radius of those techniques. A third, but much less prominent, goal has been to design robust solution methods which retain approximately optimal convergence characteristics despite the 'small' (path dependent) variations in the iteration matrices which arise automatically in nonlinear systems and in the small alterations which are part of a sequence of simulations or forecasts.

This paper is principally concerned with the robustness properties of model solution techniques. Some recent work has proposed using nonstationary extrapolations

H.M. Amman et al. (eds), Computational Economics and Econometrics, 107–119.

of the conventional iterative solution methods.[1] Nonstationary extrapolations have obvious attractions because they can be adapted to handle solution path dependent iteration matrices automatically and without interventions by the user. A leading example is the USA algorithm of Khilnani and Tse (1985). However these techniques seem to have run into considerable problems of nonconvergence, false convergence or instability, and excessive computational costs (Herceg and Cvetkovic, 1989). Moreover the existing convergence analysis appears to be incomplete, and the only experience of their performance is in small physical systems and highly simplified demonstration examples. The purpose of this note is to review the convergence characteristics of these non-stationary extrapolative procedures and to consider their likely advantages when applied to large nonlinear econometric systems.

2. First-Order Iterative Techniques

2.1. Specification:

Consider the linear equation system

$$Ay = b \tag{1}$$

where $A \in \Re^{n,n}$ is a known real matrix of order n with nonvanishing diagonal elements, and where y and b are real vectors containing the unknown and predetermined parts of each equation respectively. Stationary first-order iterative techniques

$$y^{(s+1)} = Gy^{(s)} + c\,, \quad s = 0,\, 1,\, 2,\, \ldots \tag{2}$$

with an arbitrary start $y^{(0)}$, are routinely used to construct the numerical solution to (1). They are computationally efficient, compared to (say) Newton methods, if A is a large, sparse or illconditioned matrix [Hughes Hallett and Fisher (1988)]. They are also widely used for solving nonlinear equation systems, in which case A represents the system's Jacobian matrix.

These first-order methods are based on the splitting $A = P - Q$ where P is nonsingular, and they are completely consistent with (1) when $G = P^{-1}Q$ and $c = P^{-1}b$ define the iteration matrix and forcing function [Young, 1971]. The most widely used splittings are those which produce the Jacobi, Gauss-Seidel and Successive Overrelaxation (SOR) iterative methods: respectively

$$P = D \text{ and } P = (D - L) \text{ and } P = \frac{1}{\alpha} D(I - \alpha D^{-1} L) \tag{3}$$

where D and L are matrices or order n such that

$$D_{ij} = \begin{cases} A_{ij} \text{ if } i = j\,, \\ 0 \quad \text{otherwise}\,, \end{cases} \quad \text{and } L_{ij} = \begin{cases} -A_{ij} \text{ if } i < j\,, \\ 0 \quad \text{otherwise}\,, \end{cases}$$

[1] See Khilnani and Tse (1985), Hughes Hallett (1985), or Herceg and Cvetkovic (1989).

where $U = D - A - L$ and α is a relaxation parameter to be chosen. Without loss of generality, let (1) contain the usual normalisation $D = I$, and let $B = I - A$.

The rate of convergence in (2) is often significantly increased by the one-parameter extrapolation

$$y^{(s+1)} = \lambda(Gy^{(s)} + c) + (1 - \lambda)\, y^{(s)} = Hy^{(s)} + \lambda c\,. \tag{4}$$

Two particular versions of (4) are routinely used; the Jacobi overrelaxation (JOR) method with $G = B$, and the Fast Gauss-Seidel (FGS) method based on

$$G = (I - \alpha L)^{-1}\,(\alpha U + (1 - \alpha)\, I)\,. \tag{5}$$

Fast Gauss-Seidel iterations were introduced as the Accelerated Overrelaxation (AOR) method by Hadjidimos [1978] and separately as the FGS method by Hughes Hallett [1982]. However Sisler [1973] appears to be the first to have used it as a solution technique. They represent the most general form of a first order extrapolation [Hughes Hallett, (1986)].

In a nonlinear equation system such as $y = g(y, b)$, the JOR, SOR and FGS iterations are defined as

$$y_i^{(s)} = \lambda g_i(y_i^{(s-1)} \ldots y_n^{(s-1)}) + (1 - \lambda)\, y_i^{(s-1)} \tag{6a}$$

or

$$\tilde{y}_i^{(s)} = \alpha g_i(\tilde{y}_i^{(s)} \ldots \tilde{y}_{i-1}^{(s)},\, y_{i+1}^{(s-1)} \ldots y_n^{(s-1)}) + (1 - \alpha)\, y_i^{(s-1)} \tag{6b}$$

with

$$y_i^{(s)} = \lambda \tilde{y}_i^{(s)} + (1 - \lambda)\, y^{(s-1)} \tag{6c}$$

respectively, for $i = 1 \ldots n$ and where $\lambda = 1$ in (6c) generates SOR iterates. In this case we replace G and c by $G^{(s-1)}$ and $c^{(s-1)}$ in (4). Let $B^{(s)} = \partial g/\partial y$, evaluated at $y^{(s)}$, be partitioned as before: $B^{(s)} = L^{(s)} + U^{(s)}$. Then JOR has $G^{(s-1)} = B^{(s-1)}$ and FGS has $G^{(s-1)}$ defined by (5) with $L^{(s)}$ and $U^{(s-1)}$.

2.2. Convergence Results:

It is well known that (2) converges if and only if $\rho(G) < 1$, where $\rho(\cdot)$ denotes spectral radius [Young (1971) theorem 5.1]. Given that:

Lemma 1: (4) converges for some $\lambda > 0$ if and only if $a_j < 1$, $j = 1, \ldots, n$, where G has eigenvalues $\mu_j = a_j + ib_j$ and $i = \sqrt{-1}$. (proof: Hughes Hallett, 1984.)

Lemma 2: Let $r_j^2 = (a_j - 1)^2 + b_j^2$ and $\beta = \min_j\,(\beta_j)$ where $\beta_j = -2(a_j - 1)/r_j^2$. Then under the conditions of Lemma 1, the optimal value for λ in (4) is

$$\lambda^* = \min\,(\min_k\,\{\lambda_k | \frac{1}{2}\beta_k < \lambda_k \leq \beta\},\ \frac{1}{2}\beta_{k-1}) > 0$$

where $\lambda_1 = \min_{j=1,\ldots,n}\{1 - a_j\}$ defines a_1 among the set a_j, and $\lambda_k = \min_{j>k-1}$ $\{1(a_{k-1} - a_j)/(r_j^2 - r_{k-1}^2)\}$ likewise defines a_k (for $k = 2, \ldots, n-1$). It is convergent only if $0 < \lambda < \beta$. (proof: Hughes Hallett 1984.)

LEMMA 3: $\alpha > 0$ exists small enough such that SOR in (2) is convergent if the matrix B has all eigenvalues with real parts less than unity. But it converges only if $0 < \alpha < 2$. (proof: Hughes Hallett, 1986.)

It has been possible to extend Lemma 3 only for certain special cases. The main results are:

LEMMA 4: SOR is convergent for $0 < \alpha < 2$ when A is symmetric and positive definite, irreducibly or strictly diagonally dominant, or an M-matrix. Moreover SOR and JOR are both convergent, but SOR is faster, given $\alpha = \lambda$ if $0 < \alpha < \beta$ and B is either a non-negative matrix or weakly block cyclic with $\rho(B) < 1$.[2]

The problem here is that the relationship between α and the speed of convergence, and hence the optimal or permissible values for α, are not known. Consequently the convergence (and superiority) of SOR can be established only in certain special cases, none of which normally hold for an econometric model whose only possible restriction is that B should be real. Hence the Fast Gauss-Seidel extrapolations have proved the more helpful device for accelerating Gauss-Seidel iterations.

In a nonlinear system convergence to a fixed point y^*, given an arbitrary start $y^{(0)}$ within a neighbourhood of y^*, follows if $\rho(G^*) < 1$ where G^* is $G^{(s-1)}$ evaluated at y^* [Ostrowski, 1966]. By Lemmas 1 and 3, convergence also follows for some $\lambda > 0$ and $\alpha > 0$ only if $a_j < 1$ all j, where $\mu_j = a_j + ib_j$ are roots of B^*. Similarly Lemmas 2 and 4 hold in terms of B^*.

3. Nonstationary Extrapolations and The USA Algorithm

In a nonlinear system the iteration matrix becomes solution path dependent. Thus the roots of $B^{(s)}$ and $G^{(s)}$ vary with the solution path, as do the optimal and admissible values of λ. It may be necessary, and it will generally be desirable, to reset λ at regular intervals during the iteration.

There are various ways of calculating approximately optimal λ values during the iterative process itself. These calculations should be automatic, depending only of information generated by the iterative process. They can then be programmed

[2] For detailed proofs of the results summarised in this lemma, the reader is referred to: (a) Theorem 4.3.6 of [Young, 1971] when A is symmetric and positive definite; (b) Theorem 2 of Herceg and Cvetkovic (1989) when A is irreducibly diagonally dominant and $0 < \alpha \leq 1$, and extended for $1 < \alpha < 2$ in [Hughes Hallett, 1981]; Theorem 4.5.1 of [Young, 1971] when B is nonnegative; and (d) Theorem 6.2.2 of [Young, 1971] when B is 2-cyclic and consistently ordered; (e) Theorem 2 of Herceg and Cvetkovic (1989) when A is an M-matrix. Finally, for the companion results of SOR vs. JOR, and the restriction of $0 < \alpha < \beta$, see Hughes Hallett (1981), p. 302–3.

into the solution algorithm and guide the solution path without any interventions by the user.

Perhaps the simplest of these automatic extrapolation methods is to replace λ at each step at either (6a) or (6c) with:

$$\lambda_s^* = (1 \pm [\text{abs}\,(y_i^{(s)}/y_i^{(s-2)})^{\frac{1}{2}}])^{-1} \quad \lambda_0 = \lambda_1 = 1 \tag{7}$$

the sign being positive if the iterations are cycling, negative if monotonic, and where y_i is the element which violates the convergence criterion by most [Hughes Hallett, 1985].[3] To avoid large changes in λ_s^*, (7) may be recomputed at regular intervals and smoothed; e.g.,

$$\lambda_s = w_s\,\lambda_s^* + (1 - w_s)\,\lambda_{s-1} \tag{8}$$

where w_s can be preset or varied with the iterations, and λ should be restricted to the open interval $(0, 2]$ and w_s to $[0, 1]$.

Another automatic method is the USA algorithm (Khilnani and Tse, 1985). This requires arbitrary start vectors $y^{(0)}$ and $\tilde{y}^{(1)}$ to match (6c). It also requires the error and change in error vectors

$$e_s = y^{(s)} - \tilde{y}^{(s)} \text{ and } \Delta e_s = e_s - e_{s-1} \tag{9}$$

Then λ in (6c) would be replaced by

$$\lambda_s = \frac{\gamma\,(y^{(s)} - y^{(s-1)})'\,\Delta e_s}{\|e_s\|\;\|\Delta e_s\|} \qquad \gamma \in (0, 1) \tag{10}$$

where $\|x\| = (\Sigma x_i^2)^{\frac{1}{2}}$ is the Euclidean norm. It is not clear if this algorithm should bound the values of λ_s.

There are four points to make about the USA algorithm:

(a) It is just another automatic extrapolation method for accelerating FGS iterates. Like (7) it only requires information generated by the iterative process itself and does not involve any interventions by the user (the setting of λ excepted). Even in the solution of linear equation systems, both (7) and (10) generate nonstationary FGS iterations.

(b) The USA algorithm involves a significant amount of extra computation at each step and the storage of at least 2 extra vectors. To evaluate (10) in a large system could therefore add significantly to the solution costs.

[3] If $|y_i^{(s)}/y_i^{(s-2)}| > 1$ then $y_i^{(s)}/y_i^{(s-2)}$ should be inverted in (7). The $(0, 2)$ restriction on λ_s will prevent λ_s^* exploding as we approach convergence in the monotonic iterates case, or reversing the solution path if the start values were too small. Further details and some alternatives to (7) are given in Hughes Hallett (1985).

(c) (7) and (10) reveal a marked difference in approach. A technique like (7) sets the extrapolation parameter to accelerate that variable which is (currently) slowest to converge — in fact (7) is designed to reduce the iteration matrix's dominant eigenvalue as much as possible at each step, and therefore to speed up the overall convergence rate. The USA algorithm, on the other hand, just provides an 'average' extrapolation parameter for all variables. That is less effective unless all variables converge at roughly the same rate. In economic models convergence is typically determined by one or two variables which converge slower than others. That happens naturally because of different degrees of interdependence and, more importantly, because accounting identities mean that aggregate variables converge slower (at least no faster) than their components. For example GDP converges only when all its expenditure components have converged; and similar conditions apply to government debt or the current account trade balance. In fact agents will always face a series of markets linked by Walras's Law. Thus differing convergence speeds will always arise to some extent. It might be argued that the model should be rescaled before solution, but that is not practical in most building/forecasting environments where the model is changing rapidly.

(d) The USA algorithm is not guaranteed to converge — refuting Khilnani and Tse's (1985) claim to global convergence and confirming Herceg and Cvetkovic's (1989) results. Any system implying B^* or G^* with roots whose real parts span unity will, for example, imply a divergent USA algorithm.

4. Some Empirical Experience

4.1. The Test Models

So far tests of the USA algorithm have been confined to demonstration examples of one or two equation systems.[4] In Fisher and Hughes Hallett (1987), we tested the JOR, SOR and FGS solution techniques on five of the most widely used models of the UK economy. We now extend those tests to include the USA algorithm.

Our models represent a wide variety of systems ranging from 32 to 1252 equations. Liverpool and CUBS are relatively small annual models, the latter emphasising the supply side and the former a new classical view with market clearing. The UK Treasury, NIESR and London Business School models are larger quarterly models with roots in the Keynesian income-expenditure framework and some international monetarist/financial components. They are all nonlinear and three of them contain forward-looking expectations (which, for the purposes of demonstrating single period solutions, have been fixed at arbitrary initial values).[5] These models were all available on the ESRC Macroeconomic Modelling Bureau's

[4] See Khilnani and Tse (1985) and Herceg and Cvetkovic (1989) respectively.
[5] In Fair-Taylor (1983) terminology, we are comparing type I iterations here.

User Service during 1986 and they are described in Wallis *et al.* (1986). We use the same model versions here with the exception of the UK Treasury model whose new 'slimline' version is considerably smaller (Melliss *et al.*, 1989).

4.2. COMPUTATIONAL COSTS

The USA algorithm requires more computation per step than a fixed parameter FGS algorithm because λ_s must be evaluated at every step. If we define one multiplication and one addition as a single arithmetic operation, then an operation count shows evaluating λ_s will cost $3n + 1$ operations. Both the USA and FGS algorithms involve $\Sigma n_i + 2n$ operations for their extrapolation stage, where there are n equations in the system and the ith equation contains n_i right hand side unknowns. Therefore the USA algorithm at least doubles the cost of each extrapolation step (since $n_j \geq 1$). The remaining calculations are the same in both algorithms.[6] Thus, if FGS converges S_F steps and the USA algorithm in S_u steps, the latter will only be cheaper computationally if

$$\frac{3n+1}{\Sigma n_i + 2n} + 1 \leq S_F/S_u \tag{11}$$

Table I evaluates this bound for each of our five models.[7] Given those values, we can compare the actual S_F/S_u results to see if the inequality in (11) is ever satisfied. Or we can compute the upper bound for S_u (given the actual value for S_F) such that the inequality is satisfied and compare the result to the actual value of S_u.

Table I reports the actual values of S_F for each model, expressed as the average number of FGS iterations to convergence for a sequence of solutions over T periods. Those figures imply that the USA algorithm will cost more unless it converges in 11 or 13 iterations (or less) on the Liverpool or CUBS models, or in 8, 6 or 4 iterations for the NIESR, HMT and LBS models. In fact we found the USA algorithm failed

[6] Note, in particular, that both algorithms have a parameter which has to be determined: λ in the case of FGS, and γ within λ_s in the case of the USA algorithm. The FGS parameter might be set by Lemma 2, although we have found that it is more efficient in practice to use a directed search to set λ and α simultaneously. Our procedure is as follows. In Fisher and Hughes Hallett (1987) we found that, conditional on any α value, $\alpha\lambda^* =$ constant. Thus the higher is α, the lower the conditionally optimal λ^*. Hence, if the model solves for $\alpha = \lambda = 1.0$, keep α fixed and reduce λ by .05 until a local optimum is reached. Then do a single step reducing α by .05 and increasing λ by .05 to find if there is any gain by using $\alpha \neq 1.0$. If the model does not solve at $\alpha = \lambda = 1.0$, reduce α until it does and then start the search. This procedure usually produces approximately optimal values in less than five steps and provides parameters which are robust to variations in the iteration matrix (see Figure 1 as an example). Unfortunately no simple search procedure has been proposed for γ in (10) although the interchange between Herceg and Cvetkovic (1989) and Khilnani and Tse (1989), and the extreme sensitivities revealed in Figures 1 and 2a, show that the value of this parameter is in fact crucial. A straightforward grid search within the USA algorithm will be significantly more expensive than the search for the FGS extrapolation parameters just described.

[7] This inequality actually favours the USA algorithm because it treats the evaluation of a square root as equal to an addition and because it ignores the $2nS_u$ additions in evaluating e_s and Δe_s.

TABLE I

Conditions for the USA algorithm to be computationally efficient compared to FGS (Averaged over T periods).

Model	Liverpool	CUBS	NIESR	LBS	HMT
R	1.77	1.59	1.60	1.65	1.47
S_F	20.9	21.2	13.1	7.3	10.23
S_u upper bound	11.8	13.3	8.2	4.4	6.96
T	14	14	23	36	37

where $R = \dfrac{3n+1}{2n+\Sigma n_i} + 1$

to converge (for any γ value) in three of these models, and needed 40 and 21 iterations for the CUBS and HMT models respectively (Table II).

Hence the USA algorithm fails to converge in 3 out of 5 cases, and is over three times more expensive in the other two.

How general are these results? Economic models have consistently specified from 2 to 4 right hand non-predetermined variables per equation, but very seldom more than 4. In fact the set of five models in Table II reveal an average of 3.1 explanatory non-predetermined variables per equation. Thus if the typical model has $\Sigma n_i \simeq 3n$, the ratio S_F/S_u has to be greater than 1.6 and the USA algorithm has to converge about 35–40% faster than its FGS counterpart before it can become competitive on the kind of models which are encountered in economics.

4.3. Convergence and Relative Convergence Speeds

In Table II we report the number of equations and non-zero off-diagonal elements in each model's Jacobi matrix. The table then shows the number of iterations required to converge using a standard fixed parameter FGS algorithm to solve a single time period given a government expenditure shock. That yields some gains in convergence speed over unaccelerated SOR (or Gauss-Seidel) iterations.[8] In fact reductions of up to 35% in the number of iterations were found. Furthermore FGS

[8] Our previous investigations has shown that SOR methods were usually effective with $\alpha = 1$; other α values only becoming important when the model fails to solve with $\alpha = 1$ and successive reductions in the value of α are needed to secure convergence. See Fisher and Hughes Hallett (1987).

TABLE II

The Number of Iterations to Convergence Using Different Solution Techniques.

			Fixed Parameter	Automatic	
Model	n	Σn_i	FGS(λ^*)	FGS($1-\omega$)	USA(γ)
Liverpool	32	62	64(0.6)	263(0.2)	∞
CUBS	88	270	15(0.6)	16(0.8)	40(0.75)
NIESR	199	599	12(1.0)	17(0.8)	∞
LBS	1252	3774	16(0.95)	32(0.3)	∞
HMT	397	1753	6(1.1)	8(0.5)	21(0.48)

Note: No convergence is defined at a maximum of 500 iterations.

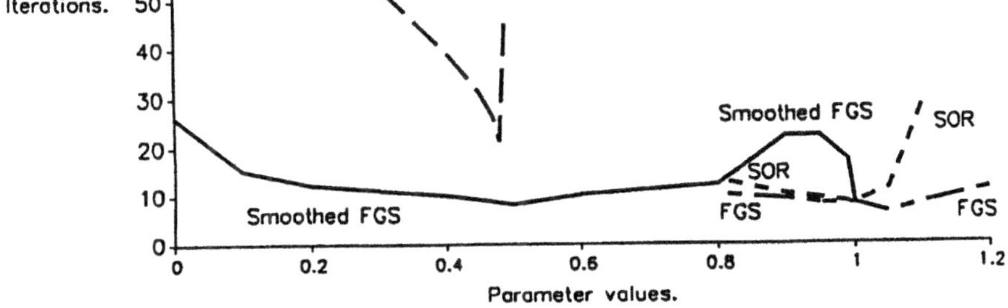

Fig. 1. Robustness to parameter choices: HMT model.

becomes more effective the harder the model is to solve. Finally FGS is robust to changes in the optimal extrapolation parameter; any λ value in the interval (0.8, 0.95) usually produces approximately the maximum convergence speed.

Fixed parameter FGS methods are therefore relatively efficient even when applied to nonlinear systems. Table II also reports the results of using an automatic FGS method with a smoothing parameter, (7) and (8), to account for the nonstationarity of the iteration matrices. In all cases, this solved the model. In 3 cases out of 5, a substantial loss is made compared to the fixed parameter approach (although the NIESR and LBS models still solve very fast in absolute terms). In the other two cases (CUBS and HMT), the automatic method gets very close to the optimal fixed parameter method.

Finally we show the results for the USA algorithm. This time we failed to find any γ value for which the algorithm would converge in 3 cases out of 5. In the remaining two cases (CUBS and HMT) the algorithm did converge, albeit three times slower than their FGS counterpart. These results were obtained using a grid search for $\gamma \in [0, 1]$, in steps of 0.05.[9] Convergence failed for some γ values because the solution path introduced invalid arithmetic (typically the log of a negative variable); but for some γ values the algorithm continued to function but failed to converge for any chosen maximum number of iterations. Moreover, the range of admissible γ values was very narrow in each case. Too small a value and convergence fails because the step size is outweighed by oscillations in λ_s. Too large a value and the algorithm fails because it steps outside feasible solution space. In either case the iterations get outside the neighbourhood for which the contraction mapping (or Lipschitz) conditions hold.

Finally there is the possibility that convergence fails because the values of a_j, defined in section 2.2 for a neighbourhood around y^*, span unity. However that does not happen here since FGS converges in each case. Nevertheless if convergence is so very sensitive to the choice of γ, we can hardly maintain that the USA algorithm is an automatic method if it requires the user to perform an accurate grid search for each of the model solutions attempted.

4.4. Robustness

The convergence problems found in Table II can be explained by examining the robustness of each method to changes in their fixed parameters. Figure 1 shows how the iteration counts vary on the HMT model as each fixed parameter is changed. The FGS and SOR searches are very flat around their optimal values. SOR is less robust than FGS, especially above its optimal parameter value. This repeats the results of Fisher and Hughes Hallett (1987). The smoothed automatic FGS parameter is also flat around its local minimum value, but there is a global minimum value where the technique collapses to fixed parameter FGS method.

[9] Khilnani and Tse (1989) relate γ to the spectral radius of $G^{(s)}$ — but that is generally an unknown, and path dependent, quantity.

The USA algorithm, in contrast, has a very pronounced minimum at $\gamma = 0.48$, whereas at $\gamma = 0.49$ the iteration costs have doubled and at $\gamma = 0.5$ the algorithm fails altogether. Hence the USA algorithm is *highly* sensitive to the choice of γ and will not be robust across a range of different forecasts or simulations. Very similar results were obtained for the convergent CUBS model solutions.

We tried various ways of avoiding the USA algorithm's apparent lack of robustness. First we noted that, although with $\gamma = 1$ the algorithm did not solve the HMT model, including an SOR damping term ($\alpha = 0.95$) allowed it to solve for 66 iterations. Thus the USA algorithm is also highly sensitive to the use of its SOR component.

Second Figure 2 shows in its upper panel the value of λ_s for various attempts to use the USA algorithm. These trials include: (a) the iteration path of λ_s when $\gamma = 0.48$; (b) the unsuccessful solution with $\gamma = 1.0$; (c) an unsuccessful solution with $\gamma = 1.0$ and λ_s restricted to the interval $(0.05, 2.0)$; and (d) the successful solution with $\gamma = 1.0$ and $\alpha = 0.95$. As we can see the USA extrapolation parameter λ_s is highly volatile. It can be compared with the lower panel of Figure 2 which shows the automatic FGS parameter unsmoothed, smoothed and the fixed FGS optimal value. Clearly, the FGS parameter is much less volatile and this procedure solves much quicker. Using the SOR parameter within the USA algorithm induces a similar degree of smoothing and hence an improved performance.

On the evidence of these results the USA algorithm needs a smoothing factor. But the more that option is used, the more we move back to a fixed parameter FGS scheme. More to the point, perhaps, is the fact that any automatic method still seems to require a grid search for a damping (or scale) factor without any gain in solution costs. Convergence without careful adjustment of those parameters by the user will be difficult to achieve for a range of different simulations or forecasts.

5. Conclusions

a) Fixed parameter FGS methods are relatively efficient, robust and easy to use when compared with other first-order extrapolations. Automatic methods are often slower and computationally unstable because of the fluctuating parameter values. They are less useful in normal circumstances, although they may be helpful for solving very difficult systems.

b) The USA algorithm is a nonstationary automatic extrapolation device within the general FGS framework. There is no extension of convergence properties therefore. Indeed the fixed parameter USA algorithm (favoured by Herceg and Cvetkovic) is none other than a standard FGS algorithm. Moreover FGS can be rewritten as a restricted form of second order iterative technique (Hughes Hallett, 1982, 1984). The implication is that it would be more effective to relax those restrictions (which will give a nonstationary method, but with converging parameters) than to supply the FGS method with nonstationary parameters.

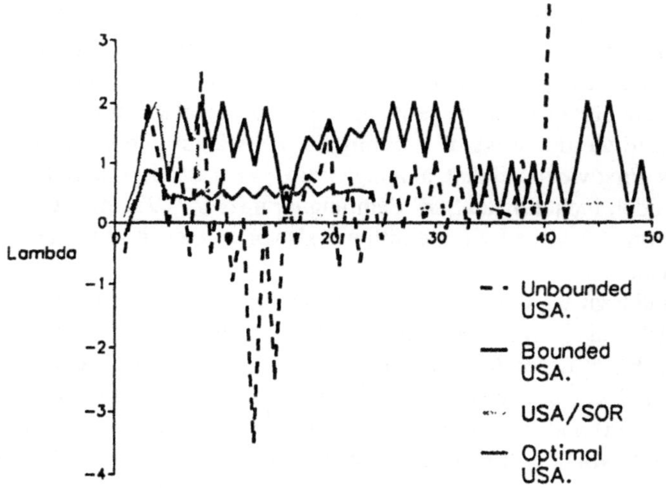

Fig. 2a. The USA algorithm.

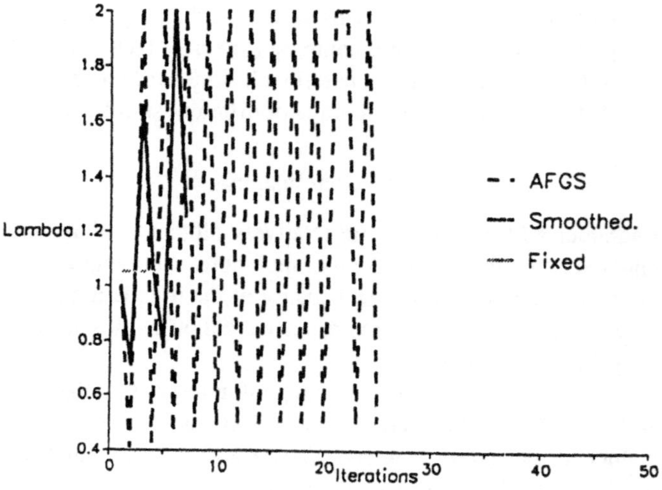

Fig. 2b. Automatic FGS.

c) In a large equation system, the USA algorithm is expensive because of the calculations involved in determining the extrapolation parameter at each step. This is a significant disadvantage as far as economic models are concerned which is not brought out by the examples in the literature. However the USA algorithm's extreme lack of robustness to its fixed parameter may prove a greater disadvantage in practice.

References

Fair, R.C. and J.B. Taylor: 1983, 'Solution and Maximum Likelihood Estimation of Dynamic Nonlinear Expectations Models', *Econometrica* **51**, 1169–1186.

Fisher, P.G. and A.J. Hughes Hallett: 1987, 'Convergence Characteristics of Iterative Techniques for Solving Econometric Models', *Oxford Bulletin of Economics and Statistics* **49**, 231–244.

Hadjidimos, A.: 1978, 'Accelerated Overrelaxation Method', *Math. Comput.* **32**, 149–157.

Herceg, D. and L. Cvetkovic: 1939, 'On the Extrapolation Method and the USA Algorithm', *Journal of Economic Dynamics and Control* **13**, 301–11.

Hughes Hallett, A.J.: 1981, 'Some Extensions and Comparisons in the Theory of Gauss-Seidel Iterative Techniques for Solving Large Equation Systems', in: E.G. Charatsis, ed., *Proceedings of the 1979 Econometric Society Meeting: Essays in Honour of Stefan Valavanis* (North-Holland, Amsterdam).

Hughes Hallett, A.J.: 1982, 'Alternative Techniques for Solving Systems of Nonlinear Equations', *J. Comput. Appl. Math.* **8**, 35–48.

Hughes Hallett, A.J.: 1984, 'Second Order Iterations with Guaranteed Convergence', *J. Comput. Appl. Math.* **10**, 285–91.

Hughes Hallett, A.J.: 1985, 'Techniques Which Accelerate the Convergence of First Order Iterations Automatically', *Lin. Algebra Appl.* **68**, 115–30.

Hughes Hallett, A.J.: 1986, 'The Convergence of Accelerated Over-relaxation Iterations', *Math. Comput.* **47**, 219–223.

Hughes Hallett, A.J. and P.G. Fisher: 1990, 'On Economic Structures and Model Solution Methods: Should Econometricians Use Newton Methods for Model Solution?', *Oxford Bulletin of Economics and Statistics* **52**, 311–324.

Khilnani, A. and E. Tse: 1985, 'A Fixed Point Algorithm with Economic Applications', *Journal of Economic Dynamics and Control* **9**, 127–37.

Khilnani, A. and E. Tse: 1989, 'A Note on the Radius of Convergence of the USA Algorithm', *Journal Economic Dynamics and Control* **13**, 313–6.

Ostrowski, A.: 1966, *Solutions to Equations and Systems of Equations*, (Academic Press, New York, 2nd Edition).

Sisler, M.: 1973, 'Über ein Zweiparametriges Iterisations-Verfahren', *Appl. Math.* **18**, 325–332.

Wallis, K.F. (ed.), M.J. Andrews, P.G. Fisher, J.A. Longbottom, and J.D. Whitley: 1986, *Models of the UK Economy: a Third Review by the ESRC Macroeconomic Modelling Bureau*, (Oxford University Press, Oxford).

Young, D.M.: 1971, *Iterative Solution of Large Linear Systems*, (Academic Press, New York).

IMPLEMENTING NO-DERIVATIVE OPTIMIZING PROCEDURES FOR OPTIMIZATION OF ECONOMETRIC MODELS

GYÖRGY BARABÁS

University of Hagen, D–5800 Hagen, West Germany *

New address: RWI, Hohenzollernstrasse 1/3, D4300 Essen 1

Abstract. The optimization algorithms used in the computer program LOS (=Large Optimizing System) are described in this paper. Not only the no-derivative but even a gradient method implemented recently are on investigation. Time consumption and robustness of the methods are reported optimizing a medium-sized non-linear econometric model on a PC.

Key words: econometric models, non-linear optimization

1. Introduction

The PC program LOS (=Large Optimizing System) has been developed at the University of Hagen in order to optimize large, non-linear, deterministic econometric models.

Originally only a no-derivative optimization method (Nelder-Mead) was used in order to optimize constrained models with very general and non-smooth objective functions. The title of this paper comes from this tradition. Recently, spring 1990, a gradient method has been also implemented.

In this paper the mathematical description of the implemented optimization algorithms is reported as well as their characteristics in a case study on a middle-sized econometric model.

If one tries to appreciate the quality of the computer program LOS it is not enough to concentrate on the objectives of this paper. There are three more classes of features that make it useful.

The one is that LOS can handle very large models of some thousand equations in the 640k memory of a PC. This feature requires compromises in the implemented procedures: They should be memory saving in code as well in work-space. Therefore e.g. all numbers are stored in single precision.

The second feature of LOS is its user friendly model handling. The model can be written by a symbolic notation and not in a particular computer code (e.g. Fortran) which is the case at many programs. The symbolic modelling allows the easy and effective error checking, altering and documenting of the model.

Thirdly, the running of traditional forcasting (simulation) is very fast. If you change something in the model (or the whole model itself) you get a forecast within

* The research was granted by the German Society of Researches (DFG, Bonn) contract No. Gr 410/6–2, applied by Prof. Dr. Josef Gruber, Ph.D.

H.M. Amman et al. (eds), Computational Economics and Econometrics, 121–135.

2–10 minutes depending on the model size (e.g. 100–1000 equations) and of course on the speed of your PC. This feature also saves you time in the preparation of a correct and error-free forecasting model.

You find further details about LOS in the User's Guide.

2. Problem Formulation

Describing the optimization problem the optimization model and the simulation model will be distinguished. The dimension of the optimization problem is reduced at a high level by this separation. The slow optimization procedures should do a small task while the simulation model is solved by a fast equation-system-solver, in our case by the Gauss-Seidel method.

The optimization problem has a very simple shape in the common formulation:

$$minimize \;\; y_g \mid y_g = g(\mathbf{I}),\; \mathbf{I} \in \mathcal{R}^m \tag{1}$$

where I is the set of control (instrument) variables, $g()$ is the objective function (mapping), y_g is the objective variable (an endogenous variable).

Any of the optimization procedure of LOS works on the way that a particular value combination will be adopted to **I**, the function $g()$ will be evaluated and y_g observed. Depending on the value of y_g a new value combination will be determined for **I** with the hope that it makes a smaller y_g. The procedure continues as long as a termination criterion has been fulfilled.

The achieved minimum is not guaranteed the global but a local minimum.

The unknowns of the optimization model are the instrument variables and the objective variable.

In case of econometric models the evaluation of $g()$ can be difficult. $g()$ can represent a non-linear vector function (equation system) and even y_g can be part of the simultaneous part of the equation system. In the letter general case $g()$ cannot be represented as a single explicite function at all.

Now we define the simulation model. This is a representation of a system of n equations which are normalized. The normalization means that each endogenous variable appears only once and alone on the left hand side of the equations:

$$\mathbf{y}^t = \mathbf{F}(\mathbf{y}^t, \mathbf{y}^t_{-\tau}, \mathbf{x}^t, \mathbf{A}^t, \mathbf{I}^t) \quad t = 1, \ldots, T \tag{2}$$

where

$\mathbf{F}$ represents n non-linear equations ($f_i \in \mathbf{F}, \forall f_i = \mathbf{F}$, $i = 1, \ldots, n$);
$\mathbf{y}$ is the vector of endogenous variables ;
$\mathbf{y}_{-\tau}$ is the vector of lagged endogenous variables;
$\mathbf{x}$ is the vector of exogenous variables;
$\mathbf{A}$ is the set of parameters;
$\mathbf{I}$ is the vector of instrument variables;
$t = 1, .., T$ denotes the time horizon.

The unknowns of the simulation model are the $n * T$ endogenous variables. The instruments are not unknown, since a) in traditional forecasting models the instruments are treated as exogenous variables with known values: $\mathbf{I} \subset \mathbf{x}$; b) in case of optimization the optimization procedure supplies always a particular value combination for **I** in order to evaluate the objective variable.

Note, the objective variable y_g is one of the endogenous variables of the simulation model: $y_g \in \mathbf{y}$.

It is worth to repeat that the minimization procedures of LOS never bother with the particular shape of $g()$. They bother only how different values of **I** influence the y_g. The evaluation of $g(\mathbf{I})$ is made by a subroutine which solves the simulation model.

What measures could the separation reduce the dimension of the optimization task? As you see in the case study, we have a model of 141 equations for 9 periods but only of between 1 and 22 instruments. It means, LOS uses the fast Gauss-Seidel solver for the 1269 $(= 9 * 141)$ simulation-model variables and the more slowly optimizing procedures only for 1–22 variables. You can define at most 50 instrument variables and some ten-thousand simulation-model variables for LOS.

It is not necessery to use the general time horizon for the instrument variables. In practical works in order to make a further reduction in the number of instrument variables you don't require to use an economical variable for each period as an instrument. For example, you can define the value-added-tax (VAT) only in each fourth period as instrument and supply the last instrument value for the next three periods. The interpretation of this definition is that the VAT is stabile through four periods. Such kind of definitons are doable by LOS.

3. No-Derivative Methods

In this section the implementation of the Nelder-Mead and of its modification the Turn-the-worst algorithms are described. You find additional features in the section *Some Remarks on the Implementation.*

3.1. The Nelder-Mead Method

The Nelder-Mead method runs in two main steps:

(i) Initial phase. A simplex is built up in the space of the instruments.
(ii) Iterational phase. The corners of the simplex (vertices) are substituted as long as they shrink to one point. This point might determine the minimum of the objective variable.

Let's start with some definitions and notations.

We say, a point $\mathbf{I}^1$ is better than an other $\mathbf{I}^2$ if $g(\mathbf{I}^1) < g(\mathbf{I}^2)$, since we search for minimum.

m is the dimension of the optimization task.

Upper index enumerates a point. Lower index enumerates a coordinate of a point: $I_i^j \mid i \in 1,2,\ldots,m;\ j \in 0,1,\ldots,m$; denotes the i-th coordinate of the j-th point.

3.1.1. *The initial phase*

A simplex will be set up in the initial phase. The simplex lays in the instrument space. We are interested only in the *m+1* corner points (vertices) of the general simplex in the *m* dimensional space. Each point consist of m coordinates.

The *m+1* points are constructed in the following way. One point ($\mathbf{I}^0$) is fixed. It contains the values of the instrument variables from the data file of LOS. They are the actual or last available values. If a value is not presented, zero is asssumed.

At the construction of the other points the $delta_i$ parameters defined in the control file of LOS play a role. Each point differes only in one coordinate from $\mathbf{I}^0$. The i-th coordinate of the i-th point ($i = 1,\ldots,m$) is determined in the following way:

$$\begin{array}{lll} I_i^i := I_i^0 * (1 \pm delta_i) & \text{if } I_i^0 \neq 0 \\ I_i^i := \pm delta_i & \text{if } I_i^0 = 0 \end{array} \tag{3}$$

As you see, both directions might be tested. If you get already by the $+delta_i$ a better point than $\mathbf{I}^0$, it is accepted. In the other case the point $\mathbf{I}'^i$ calculated by $-delta_i$ is also determined and the objective value upon $\mathbf{I}'^i$ is evaluated. The better from $\mathbf{I}^i$ and $\mathbf{I}'^i$ is chosen for $\mathbf{I}^i$.

3.1.2. *The iterational phase*

In this phase the vertices of the simplex are substituted by new points through transformations. The substitution produces either better points or points which form a smaller simplex. The aim is to shrink the simplex to one point which determines the minimal value of the objective variable. The condition whether despite the small differences among the points they are accepted as one point is defined by the `KONV` parameter in the `GOAL` command in the control file of LOS.

The Nelder-Mead method operates by the following transformations:
- reflexion,
- expansion,
- partial inside contraction,
- partial outside contraction,
- total contraction.

Each iteration begins with the reflexion. The reflexion can be followed by expansion, partial inside- or outside contraction or by a new iteration. After the partial inside- and outside contraction a total contraction might come.

The algorithm is the following:

Step 0. Start a new iteration.

- The points of the simplex are sorted. At sorting the worst point ($\mathbf{I}^w$), the second worst point ($\mathbf{I}^{w2}$) and the best point ($\mathbf{I}^b$) will be distinguished. Among the same good points the point with the lowest index is marked as the best point $\mathbf{I}^b$. Among the same bad points the point with the highest index is marked as the worst point $\mathbf{I}^w$. The second worst point $\mathbf{I}^{w2}$ is selected the same way as the worst point while the worst point is omitted.
- The termination criterion (wether the simplex has shrunk to almost one point) is checked. If it matches, the iterational phase terminates. Relate to *Repeated simplex setting* in section *Some Remarks on the Implementation.*

Step 1. Reflexion.

- The worst point will be reflected through the weighting point. The weighting point ($\mathbf{O}$) is the average of all points with the exception of the worst point:

$$O_i := \frac{\sum_{\forall j,\, j \neq w} I_i^j}{m}\ ,\ i = 1, \ldots, m \tag{4}$$

The reflected point is $\mathbf{R}$. The value of *alfa* can be defined in the control file, by default 1.05:

$$\mathbf{R} := \mathbf{O} + alfa * (\mathbf{O} - \mathbf{I}^w) \tag{5}$$

- If $\mathbf{R}$ is better than the till now best $\mathbf{I}^b$, goto step 2 (expansion).
- If $\mathbf{R}$ is better than the second worst point $\mathbf{I}^{w2}$, substitute the worst point by $\mathbf{R}$ and goto step 0 (start a new iteration). It is important to test against the second worst point and not against the worst point because infinite loops might arise otherwise.
- If $\mathbf{R}$ is better than $\mathbf{I}^w$, goto step 3 (partial outside contraction).
- Else (if $\mathbf{R}$ is worse than or the same bad as $\mathbf{I}^w$) goto step 4 (partial inside contraction).

Step 2. Expansion.

- The weighting point will be reflected through $\mathbf{R}$. The value of $gamma$ can be defined in the control file, by default 2.05:

$$\mathbf{E} := \mathbf{R} + gamma * (\mathbf{R} - \mathbf{O}) \tag{6}$$

- The worst point is substituted by the better from $\mathbf{E}$ and $\mathbf{R}$. Goto step 0 (start a new iteration).

Step 3. Partial outside contraction.

- The value of $beta$ can be defined in the control file, by default 0.45 The transformation is:

$$\mathbf{C} := \mathbf{O} + beta * (\mathbf{R} - \mathbf{O}) \tag{7}$$

- Goto step 5.

Step 4. Partial inside contraction.

- $$\mathbf{C} := \mathbf{O} + beta * (\mathbf{I}^w - \mathbf{O}) \tag{8}$$

  The $beta$ is the same as in the preceeding step.

- Goto step 5.

Step 5. Conditions.

- If $\mathbf{C}$ is better than the second worst point $\mathbf{I}^{w2}$, substitute the worst point by $\mathbf{C}$. Goto 0 (start a new iteration).
- Else goto step 6.

Step 6. Total contraction.

- Contract all points except of the worst point towards the best point. The best point remains unchanged. Reflect the worst point through the best point but nearer than the original:

$$\text{for } j = 0, \ldots, m \text{ and } i = 1, \ldots, m$$

$$\begin{array}{ll} I_i^j := 0.5 * (I_i^b + I_i^j) & \text{if } j \neq b \text{ and } j \neq w \\ I_i^w := I_i^b + 0.3 * (I_i^b - I_i^w) & \text{if } j = w \end{array} \tag{9}$$

- Goto step 0 (start a new iteration).

Exhibit 1 demonstrates the different transformations in the two dimensional case with the exception of the total contraction you find in exhibit 2.

At last I sum up some features of the implementation.

(i) Stressing the role of the second worst point is important.
(ii) The reflection of the worst point at the total contraction.
(iii) The comparison of the points to the *equal to* condition is treated in the publications not uniquely: Sometimes such evaluation belongs to the one condition, sometimes to the complementary one.

The statements (i) and (iii) reflect the leading principle of the Nelder-Mead method that a new point should be either better than the earlier or form a smaller simplex in order to avoid infinite flipping of the points.

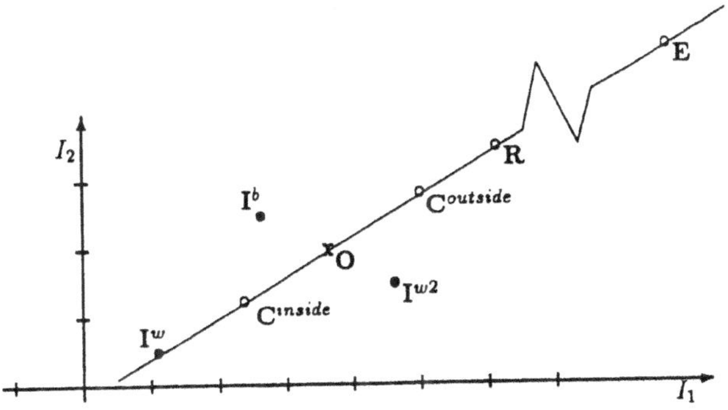

Exhibit 1. The worst point $\mathbf{I}^w$ of the original simplex $(\mathbf{I}^b, \mathbf{I}^{w2}, \mathbf{I}^w)$ will be substituted either by $\mathbf{E}$, $\mathbf{R}$, $\mathbf{C}^{inside}$ or $\mathbf{C}^{outside}$ related to the transformations expansion, reflexion and partial contractions respectively. $\mathbf{O}$ is the weighting point.

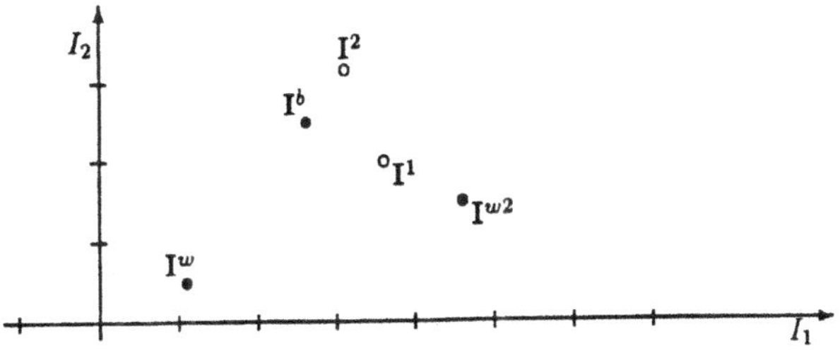

Exhibit 2. Total contraction in the two dimensional case. The simplex $(\mathbf{I}^b, \mathbf{I}^{w2}, \mathbf{I}^w)$ is reformatted to $(\mathbf{I}^b, \mathbf{I}^1, \mathbf{I}^2)$.

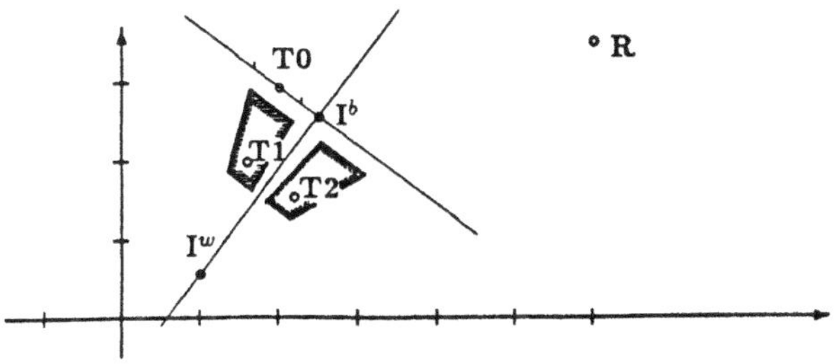

Exhibit 3. Any of the shaded points could be randomly selected as **T1** and **T2** in the Turn transformation. The coordinate-system is a hyper-plane on the $(\mathbf{I}^b, \mathbf{I}^w, \mathbf{R})$ points.

3.2. The Turn-the-worst Method

The Turn-the-worst method works similar to the Nelder-Mead. The only difference is that instead of the *inside contraction* transformation a new transformation the turning (revolving) is implemented.

The main idea of the turning transformation is as follows: I found in the early development phase of LOS that the Nelder-Mead method couldn't be adjusted at 'steep walls' (e.g. constraint by penalty function) but converged against the wall and stopped in a bad position. By the random 'Turn' transformation I tried to force the simplex in a position to find the way along the steep wall. In earlier implementations it worked better than the Nelder-Mead method but it hasn't superior any longer. See therefore the section *Some Remarks on the Implementation.*

The steps to compute the turning transformation are the followings, they substitute step 4 in the Nelder-Mead procedure:

- If the distance ($\mathbf{I}^b$, $\mathbf{I}^w$) is very small, goto step 6 of the Nelder-Mead method (total contraction) in order to avoid computational errors (underflow).
- Compute the point **T1** in the following way: The point **T0** is defined on the line laid through $\mathbf{I}^b$ by 90° against the ($\mathbf{I}^b$, $\mathbf{I}^w$) line on the hyperplane defined by the ($\mathbf{I}^b$, $\mathbf{I}^w$, **R**) points. The distance of **T0** to $\mathbf{I}^b$ is randomly taken between $0.1 * d$ and $0.4 * d$ where d denotes the length of the ($\mathbf{I}^b$, $\mathbf{I}^w$) section. A next line is set up through (**T0**, $\mathbf{I}^w$). **T1** is defined on this line randomly between $0.1{*}\mathbf{T0}\vec{,}\mathbf{I}^w$ and $0.5{*}(\mathbf{T0}\vec{,}\mathbf{I}^w)$.
  Exhibit 3 makes the calculation clearer. The parameters 0.1, 0.4, 0.5 have been chosen heuristically after some trials.
- If **T1** is better than the till now best $\mathbf{I}^b$, assign **R**:=**T1** and **O**:=$\mathbf{I}^b$. Goto step 2 of the Nelder-Mead method: expand $\mathbf{I}^b$ through **T1**.
- If **T1** is better than the second worst $\mathbf{I}^{w2}$, assign $\mathbf{I}^w$:=**T1**. Goto step 0 (start a new iteration).
- Determine **T2** symmetrically of **T1** through the ($\mathbf{I}^w$, $\mathbf{I}^b$) line.
- If **T2** is better than the till now best $\mathbf{I}^b$, assign **R**:=**T2** and **O**:=$\mathbf{I}^b$. Goto step 2 of the Nelder-Mead method: expand $\mathbf{I}^b$ through **T2**.
- If **T2** is better than the second worst $\mathbf{I}^{w2}$, assign $\mathbf{I}^w$:=**T2**. Goto step 0 (start a new iteration).
- Substitute the worst point by the better point from **T1** and **T2**. Make a total contraction transformation (as in step 6) but let the worst point just substituted unchanged.
  Goto step 0 (start a new iteration).

4. The Gradient Method

The steepest descent method has been implemented recently to LOS. However, this method has been reported as a slow one but still easy to implement. Easy implementation is important because there is a big chance at huge programs like LOS to implement even a bug by changes. Furthermore the steepest descent method requires a small amount of memory, only the gradient vector should be stored. These are the reasons while I chose the steepest descent as the first gradient method to implement.

The steepest descent method works by the following steps:

(i) Numerical computation of the gradient vector.

(ii) Normalization of the gradient vector.

(iii) Search on the gradient direction for the smallest value as long as the convergent criterion is fulfilled.

These steps are repeated by setting up a new simplex as it is described in the section *Some Remarks on the Implementation.*

The steps are the followings in details:

4.1. COMPUTATION OF THE GRADIENT

A simplex is set up the same way as written for the initial phase of the Nelder-Mead method with the difference that instead of the single $delta_i$ the product $delta_i * step\_length$ is used to determine the side length of the simplex. The value of $step\_length$ is defined in the control file of LOS. The gradient vector $\mathbf{G}$ is calculated by:

$$G_i := \frac{g(\mathbf{I}^i) - g(\mathbf{I}^0)}{\mathbf{I}_i^0 - \mathbf{I}_i^i}, \; i = 1, \ldots, m \qquad (10)$$

where $g()$ is the objective function. The gradient is two-sided since the better directions are chosen at the simplex-setting. If still $g(\mathbf{I}^i) > g(\mathbf{I}^0)$ then the i-th gradient element is zero.

Special cases are also concerned, i.e. if one of the $\mathbf{I}^i$ or $\mathbf{I}^0$ lays in the infeasible region.

4.2. NORMALIZATION

A normalized gradient vector $\mathbf{G}^N$ should be determined if any of the element of the gradient spans over the corresponding side-length of the simplex which would be set up in the initial phase of the Nelder-Mead method, i.e. without the $step\_length$ parameter. Varying the $delta_i$ parameter you can prescribe where the line search on the gradient direction should start.

4.3. LINE SEARCH

An interval should be found which includes the minimum. Thereafter the boundaries of the interval should be moved closer and closer to each other until they meet at the minimum.

As first point of investigation $\mathbf{I}^1 := \mathbf{I}^0 + \mathbf{G}^N$ is chosen. If $\mathbf{I}^1$ is worse than $\mathbf{I}^0$, the minimum lays between them. If $\mathbf{I}^1$ is better than $\mathbf{I}^0$, expansions are done by

$$\mathbf{I}^1 := \mathbf{I}^1 + (\mathbf{I}^1 - \mathbf{I}^0) \tag{11}$$

as long as the last $\mathbf{I}^1$ in the series becomes worse than the preceeding $\mathbf{I}^1$. The minimum should lay between the last point and two points before (i.e. the point before the best $\mathbf{I}^1$ in the series).

Denote the interval in which the minimum lays by $(\mathbf{I}^2, \mathbf{I}^3)$. Shrunk the interval in order to include always the best point. Do as long as the two points become the same by a tolerance criterion.

5. Some Remarks on the Implementation

There are several features in the implementation that help to gain the more correct and quick result of the optimization. This section reports also the changes that made the Turn-the-worst method less important to the Nelder-Mead.

5.1. Repeated Simplex-Setting

The optimization procedures start more times but at least two by setting up a new simplex. The minimum is accepted only if the last two starts give the same solution.

The initial point of a new simplex is the best solution point derived from the last solution. The two last results are accepted as the same if both the function values and the shrunk simplexes are the same with some tolerance talked in the User's Guide. From the second simplex the $delta_i$ parameters are set to the tenth of the originals.

In case of the gradient method the set up of a new simplex is trivial because a new gradient direction should be probed. Anyway, if the gradient vector is zero in all elements, a new simplex launches the Turn-the-worst method using the original side length ($delta_i/10$ without the $step\_length$ parameter) in order to make a more accurate calculation and to find a possible region where the gradient method doesn't operate. If its solution differes from the last, again the gradient method tries to come in use.

5.2. Saved Best Point

Each evaluation of the simulation model (Gauss-Seidel) begins from the till now best solution which has been always saved.

5.3. Decreased Tolerance

If a new value of the objective variable is better than the till now best or even worse at most by the convergent criterion of the Gauss-Seidel method, the evaluation of the equation system will be repeated more accurately (if it contains at least one simultaneous block). The new convergence criterion of the Gauss-Seidel procedure for the repetation is set to the tenth of the original.

5.4. RE-EVALUATION

In case of the total contraction in the no-derivative methods the till now best solution will be re-evaluated. This new value becomes the new best objective value although it can be also worse (i.e. larger) than the earlier.

This option makes a good improvement for the optimization if the new evaluation is really better but a deterioration if the solution is not precise.

As you see, the last three points of this section deal with the overcoming of the unaccuracy of the Gauss-Seidel method. The error of calculation is locally at the Gauss-Seidel method. The method yields sometimes larger, sometimes smaller values in the result than the (unknown) real solution would.

A seemingly good objective value can blockade the searching to the really good direction. If the re-evaluation cuts this constraint by a more realistic higher value, the way is opened for further search.

If the re-evaluation is unaccurate, it can set up new blockades and even help for moving away from the already found pritty good solution.

5.5. NELDER-MEAD VERSUS TURN

In LOS issue 1.32 the Turn-the-worst method seems not to be better than the Nelder-Mead but still slower. In the LOS issue 1.11 it was not the fact. That time the Nelder-Mead method didn't yield such a robust solution as now you see in the section *Case Study*. The Turn-the-worst method worked almost as good as now but a bit slower.

Which are the changes between issue 1.11 and 1.32 that effected the optimization characteristics? The decreased tolerance and re-evaluation discussed above in this section didn't exist earlier. The simplex was set up also another way: The vertices differed not only in one coordinate I_i^i from $\mathbf{I}^0$. All other coordinates I_j^i were also varied randomly from I_j^0 by a factor zero to $delta_j$.

6. Case Study

The three different optimizing methods are compared in this section relating to a middle-sized econometric model. The time consumption of the methods is investigated using different number of instruments and changing the tolerance criterion of the calculations.

Exhibit 4 Accuracy and time consumption of the different optimization methods. See description in the text.

Line No.	Opti. meth.	Minu-tes	No.of calls	Objective EWA	Instruments q1	q2	q3	q4
1	G3	6.7	63	23645.529	2.004	-	-	-
2	N2	5.6	70	23645.531	2.002	-	-	-
3	T2	5.7	67	23645.537	2.001	-	-	-
4	G5	19	223	23691.707	2.006	2.002	2.018	2.002
5	N3	42	452	23691.775	2.002	2.001	2.003	2.001
6	T3	48	565	23691.729	2.021	2.001	2.002	2.001
7	G11	85	1127	23736.441	2.003	2.000	2.005	2.005
8					2.008	2.000	2.006	2.001
9	N6	175	1999	23736.947	2.000	2.005	2.003	2.001
10					2.002	2.000	2.001	2.000
11	T5	194	2040	23736.398	2.001	2.008	2.014	2.000
12					2.000	2.000	2.027	2.000
13	G27	360	6251	24361.410	2.034	2.007	2.002	2.008
14					2.004	2.032	2.007	2.045
15					3.001	3.001	3.000	3.002
16					3.014	3.004	-	-
17					19.997	19.530	19.998	20.000
18					19.989	19.999	19.998	19.980
19	N18	1999	26475	24364.262	2.005	2.001	2.001	2.002
20					2.001	2.001	2.001	2.001
21					3.000	3.011	3.000	3.002
22					3.000	3.002	-	-
23					19.988	19.991	19.988	19.991
24					19.993	19.988	19.987	19.994
25	T18	1879	27610	24364.141	2.008	2.003	2.005	2.015
26					2.005	2.004	2.006	2.011
27					3.000	3.008	3.003	3.008
28					3.011	3.006	-	-
29					19.997	19.999	20.000	19.999
30					19.998	19.984	19.997	19.949

6.1. THE MODEL

First let us summarize briefly the econometric model in use. The model is a quarterly business cycle model of West-Germany, built in the Institute of Economic Research (RWI, Essen). It consists of 141 mostly non-linear equations. 58 equations are simultaneous. The number of behavionar equations is 41.

For the optimization the time horizon is set from the 4th quarter 1988 to the 4th quarter 1990. These nine periods make a system of 1269 equations. The objective variable `EWA` denotes the number of employees (in thousand) in the 4th period of 1990 which we want to maximize. The instrument variables are the short-term interest rate `ZINSK`, the long term interest rate `ZINSL` and the public investments `IBST`.

Explicit restrictions in the model have been defined for the instrument variables since economically unrealistic values would be gained for the optimum. `ZINSK` can be between (2,10), `ZINSL` (3,12) and `IBST` (5,20). The exogenous baseline data of `ZINSK` and `ZINSL` lay between 6 and 8 per cent, of `IBST` between 7 and 14 billion Deutsche Mark.

The computations have been made by LOS issue 1.32 on an 80386sx PC with coprocessor, clock rate 16 MHz.

6.2. OPTIMIZATION: THE BASELINE CASE

In this subsection the optimization on the above presented model are showed. The number of instrument variables changes from one to twenty-two. The baseline case presents pritty accurate results, attention should be taken to the calculation time. The next subsection deals with the less accurate cases, the parameters of precision setting are discussed there.

Exhibit 4 showes comprehensive data about the maximization of `EWA`. The meaning of the different columns of the table are discussed in turn:

Line No. counts the lines of the table.

Opti.meth. shows the optimizing method in use: 'N' stands for Nelder-Mead, 'T' for Turn-the-worst, 'G' for gradient. The number following the letter says how many times a simplex has been set up.

Minutes contains the time consumption of the computations in minutes. This figure doesn't include the time for reading the model and the data, writing the reports which takes less than 40 seconds in all cases.

No.of calls is the number, how much the the simulation-model solver (Gauss-Seidel) has been called.

Objective EWA gives the value of the objective variable in the 4th quartal of 1990, which we want to maximize (employees in thousand).

Instrument values q1 ... q4 are the computed values of the instrument variables in use. They refer to different quarters of 1989 and 1990.

In lines 1–12 the short term interest rate (ZINSK) is the only instrument variable through one, four and eight quarters.

In lines 13–30 the long term interest rate (ZINSL) is defined additionally as instrument from 1st quarter 1989 through 2nd quarter 1990 as well as the public investment through eight periods. They make 22 instruments together. The long term interest rate in the two last quarters of 1990 doesn't influence EWA in the 4th quarter 1990 by the model.

For the evaluation of the table the first thing is to decide which results are correct. We don't know analitycally the solution but have made many experiences by the model. All these practicies show, that the the maximum of EWA is taken if ZINSK and ZINSL are on their minimum (2 and 3) and IBST on its maximum (20). Concerning this information the results look pritty good, independently of the optimization method.

The only higher difference is in case of the 22 instruments, the gradient method calculated 2800 employees less then the no-derivative methods. Anyway, this difference doesn't seem to be important concerning the economic meaning.

If you look at the time used for computation, the gradient method uses half the minutes in case of 4–8 instruments and fifth at 22 instruments compared to the no-derivative methods. Nevertheless, I could present examples where the no-derivative methods converge faster or uniquely at all.

6.3. Increasing the Tolerance

An interesting question is how the tolerance criterion influences the robustness and time consumption of the methods. Exhibit 5 summarizes it using the eight instruments the same way as exhibit 4 (i.e. the baseline) in lines 7–12.

You can change the convergence criterion of the Gauss-Seidel solution (GKONV) and of the optimization (IKONV). The last criterion says how close the vertices of the simplex should be to each other in order to accept the simplex as one point.

In the baseline optimization both tolerances were set to 10^{-4}. If GKONV is set to 10^{-2} (weak) and IKONV to 10^{-4} (strong) the procedures won't stop. The case discussed at *Re-evaluation* in section *Some Remarks on the Implementation* occures: In consequence of the inaccurate calculation the best point of the simplex is sliding away and towards the real solution and the simplex will never accepted as one point because of the strong IKONV criterion. Two further tolerance combinations are reported in exhibit 5.

There are three further parameters of LOS which are important to know in order to reproduce the results. In all experiences the allowed maximum number of Gauss-Seidel iterations (during one evaluation) was set to 40. The $delta_i$ parameters were set on their default 0.5 . The $step\_length$ parameter for the gradient method had the value 1.

Exhibit 5. Optimization by eight instrument variables and with different tolerance criteria. EWA diff. shows the deterioration of the objective value vs. baseline of the same method.

	baseline			1st alternation			2nd alternation		
GKONV IKONV	10^{-4} 10^{-4}			10^{-3} 10^{-4}			10^{-2} 10^{-3}		
Method in use	G	N	T	G	N	T	G	N	T
Minutes	85	175	194	42	135	133	52	78	62
No. of calls	1127	1999	2040	831	2280	2415	1214	1642	1227
EWA diff.	0	0	0	0.3	0.6	0.02	1.2	11.0	11.3

Exhibit 5 shows in the 1st alternation that the no-derivative methods require two-third, the gradient method only half of the baseline time without loss of accuracy. In the 2nd alternation the no-derivative methods require half and third time of the baseline calculation but the value of EWA deteriorated by eleven-thousand. The interest rate takes in some periods a value over 4 percent (not presented in the table).

All the results of exhibit 5 propose for the practical jobs that you could start the computation by weaker tolerances to have a first quick guess about the model.

Anyway, there is no need to wait the whole time of computation at the first time you optimize a model since you can monitor the values of selected variables by LOS and interrupt the computation if any economically unacceptable result appears. In that case you can define e.g. new constraints and start again. You should choose stronger tolerance if you wish to get a numerically more precise solution. In advance you cannot mention exactly which tolerance is weak or strong because it depends also on the particular model.

For possible future improvements the case study suggests that increasing the speed and accuracy of the computation of the simulation model could do a lot for LOS.

References

Avriel, Mordecai: *Nonlinear programming. Analysis and methods*. Prentice-Hall, 1976.

Barabás, György: *User's Guide of LOS 1.32*, University of Hagen, 1990.

Fletcher, R.: *Practical Methods of Optimization*. J. Wiley & Sons, 1987.

Nemhauser, G. L., Rinnooy Kan, A. H.G, Todd, M. J. (eds.): *Optimization*. Elsevier Science Publishers B.K., 1989.

Schwefel, Hans-Paul: *Numerical optimization of computer models*. J. Wiley & Sons, 1981.

INFORMATION IN A STACKELBERG GAME BETWEEN TWO PLAYERS HOLDING DIFFERENT THEORETICAL VIEWS: SOLUTION CONCEPTS AND AN ILLUSTRATION

HENRIËTTE M. PRAST

University of Amsterdam, Faculty of Economics and Econometrics, Department of Macroeconomics, Jodenbreestraat 23 1011 NH Amsterdam

Abstract

This paper analyzes fiscal policy and wage determination as a Stackelberg game between a government and a single trade union. It is assumed that the players believe in different economic theories, which implies that their ideas about the probable effects of fiscal policy measures and wage claims differ.

The difference between the players' views is represented by differences in the parameter values of their model. Adaptations of the Stackelberg solution concept dependent on the information of the leader about the follower's model are introduced. Assuming there is a "true" model, numerical simulations are carried out in order to investigate the implications of the assumption of conflicting views for the strategies that come about as well as for macroeconomic performance and the values of the welfare loss functions of the players. It is concluded, that there may be cases in which the follower has an incentive to keep the leader unaware of the fact that their models are different.

1. Introduction

The stagflation in European economies in the seventies and the persistence of unemployment after the shift to monetarist-oriented policies in the eighties have been explained by applying game-theoretic concepts to the interaction between employment policies and wage formation (Calmfors 1982, Calmfors and Horn 1986, Driffill 1985, Gylfason and Lindbeck 1986, Hersoug 1985). The government and a trade union are seen as policy makers, each controlling an instrument and optimizing an objective function. Most of the discussion focuses on suboptimal outcomes (in terms of employment and price stability) due to the anticipation by the trade union of a government policy rule for unemployment. Efforts to reduce unemployment by expanding public sector employment are shown to "crowd-out" private sector employment and to increase wage inflation by changing the position and slope of the labor demand schedule.

The economic outcomes of the interaction between macroeconomic policy and wage determination can be analyzed as a game between the government and the

Hans M. Amman et al. (eds), Computational Economics and Econometrics, 137-155.

trade union and are influenced by:

i) the views of the players on the working of the economy
ii) the institutional characteristics of the game
iii) the targets of the players

In the literature on *international policy coordination* attention has been paid to model uncertainty (Holtham and Hughes Hallett 1986, Ghosh 1986, Ghosh and Masson 1988, Frankel and Rockett 1988, Frankel 1988) and its effect on welfare in Nash noncooperative and in cooperative solutions. In most applications of game theory to the interaction between fiscal policy and wage determination, however, it is implicitly assumed that in choosing their strategies the players use one and the same economic model. This implies, that – unless they have different information in the form of forecasts and the like – their ideas about the probable effects of policy measures coincide. Moreover, their opinions on the working of the economy are assumed to be correct. Consequently, in these applications no attention is paid to aspect i) (Pohjola, 1986). An exception is found in Gylfason and Lindbeck (1986), who in their game-theoretic analysis of the interaction between government spending and wage formation pay attention to the fact that the behavior of the players depends on both their preferences and their views. However, the *interaction* between the views of the players is not analyzed there.

In this paper we study a Stackelberg game between two players, a government and a single trade union, whose ideas about the working of the economy differ. This assumption implies that, first, the difference in views between the players may be responsible for sub-optimal outcomes. Second, in the context of a Stackelberg game, the multiple-model approach creates an additional source of information which may or may not be available to the leader: the model used by the other player. Thus various Stackelberg equilibria can be distinguished, dependent on the leader's knowledge of the follower's model.

Although in our opinion it is the *one-model* assumption that needs justification, we shall mention a few reasons why introducing a *multiple-model* approach is required. Firstly, in economic theory various schools of thought coexist, each having its own opinions as to what are the essential characteristics of an economic system. The "rules versus discretion" discussion regards not the ultimate targets of economic policy, but the best way how to reach them; and this has to do with different views on how the economy works. Secondly, debates between institutions that are concerned, in one way or another, with economic policy-making – political parties, trade unions – on the appropriate economic policy measures to be taken often can be traced back not so much to different objectives as to disagreement on how to realize them. In the third place, different institutions involved in policy-making or forecasting – the monetary authority, research institutes – all use their own econometric model, based on specific opinions about the working of the economy. As empirical tests are often not conclusive, there is no reason to assume that, as more sophisticated techniques become available, a convergence in opinions will come about.[1] Not all economic variables can be measured unambiguously. The various explanations that are given to the observed concurrence of high unemployment and rising wages illustrate this point. Those who, on the basis of their economic "belief", advocate government intervention explain this phenomenon in terms of hysteresis, whereas those who on the ground of their economic "vision" support laissez faire have introduced a natural rate of unemployment.

The purpose of this paper is to highlight some implications of the multiple-model assumption for the game-theoretic analysis of fiscal policy and wage determination. The paper is constructed as follows. Section 2 discusses the implications of the multiple-model assumption for game-theoretic solution concepts. In section 3 the players, their models – "perceived constraints" – and their welfare loss functions as well as the institutional framework in which they operate are introduced. In section 4 the optimal strategies that result for the solution concepts introduced in section 2 are derived. Section 5 presents the results of some numerical simulations. In that section a model that correctly describes the working of the economy is introduced. This enables us to assess the economic performance and the welfare loss resulting from the behavior of the players.

2. Different Views in Economic Games: Solution Concepts

Obviously, the pair of strategies, and thus the economic performance, that will come about in a multiple model game is different from that in the one-model situation. This applies to both Nash and Stackelberg games and it may thus be worthwhile to study the implications of multiple-modelling in all types of games. As far as *solution concepts* are concerned, however, the multiple-model assumption has implications mainly for Stackelberg games. The Stackelberg solution concept assumes that one of the players, the leader, in choosing his optimal strategy takes account of the rational reaction of the follower. The follower arrives at this reaction function by optimizing subject to his constraint (model), taking the instrument of the leader as given. Knowledge of the followers reaction function by the leader presupposes that the leader knows both the model and the objective function of the follower. As in the standard applications of game theory to economics the Stackelberg leader and the follower(s) adhere to one and the same model, the question whether the leader knows the followers model need not be asked. If it is assumed, however, that the players believe in different economic theories and thus use different economic

[1] Frankel and Rockett observe that even if convergence is assumed to take place it occurs at a speed sufficiently low to prevent policy makers in reality from reaching agreement on the true model.

models when choosing their strategies, information regarding the followers model becomes relevant for the leader. The leaders perception of the followers reaction function shall differ dependent on what information about the follower's model is available to him. Actually, the same applies to the information the leader has regarding the objective function of the follower(s). If the leader makes mistakes here, his perception of the followers reaction function is wrong even if he perfectly knows the followers economic view. However, we shall assume in this study that the players have full knowledge about each others objectives. This implies that we distinguish between various Stackelberg equilibria according to the information regarding the economic *view* of the follower that is assumed to be available to the leader.

First, the leader may know the model used by the follower; though he believes the model to be wrong, he uses it when deriving the follower's "rational" reaction. We shall call this the *"informed leader" solution* (notice, that informed here applies to the information the leader has about the follower's model, leaving open the question what knowledge he has about economic reality). *Secondly*, it may be that the leader is not aware of the fact that the follower has different opinions about the working of the economy. In this case, the followers reaction as perceived by the leader results from optimizing subject to the leader's constraint. We label this the *"ignorant leader" solution*, stressing that "ignorant" here is related merely to the lack of information on the follower's model, not to whether or not the leader's model corresponds to reality. In our opinion it may be quite realistic to assume that the leader does not know that the follower uses a different model. The mere fact that in economic studies that analyze the interaction between economic players as a game the possibility of different models is neglected gives support to the idea that the players themselves may not be aware of the differences between their own model and that of the other player(s).

In the *third* place it may be that the leader, though being aware that the follower uses a model different from his own, does not know this model. Various possible solution concepts can be distinguished here. For example, instead of guessing what the followers' reaction function is, the leader may take the follower's strategy as given. The game then turns into a Nash game, because the leader knows that he does not dispose of the information necessary to perform his leadership role. Thus, where in a one-model analysis conversion of a Stackelberg game into a Nash game may occur due to the time inconsistency problem, in the multiple-model case it may also result from an information problem. If the leader, however, is able to change his strategy during the course of the game, he may get to learn the follower's model from observing the followers behavior. The game then reconverts into a Stackelberg game as soon as the potential leader has gathered enough information regarding the followers model and thus reaction function. In this paper we do not go further into this and we shall concentrate on the informed-leader and the ignorant-leader solutions.

The role of information in a multiple-model Stackelberg game and the distinction between an informed leader solution and an ignorant leader solution has another, in our view important aspect. As is well-known, in the standard one-model Stackelberg game time-inconsistency may arise because there is an incentive for the *leader* to cheat, i.e. to renege on his announced strategy once this has had the desired effect on the follower's behavior. In the multiple-model Stackelberg game there may be another type of cheating, i.e. the *follower* may have an incentive to keep the leader ignorant of their model differences. This is because it may very well be that the welfare loss for the follower is higher an the informed-leader case. If the follower is aware of this, then it pays for him to keep the leader unaware of the fact that their models differ. We shall turn to this in section 5.

3. A Multiple-model Example

In this section we introduce a multiple-model Stackelberg game describing the interaction between budgetary policy and wage determination. We assume a single trade union to bargain with the employers over the money wage for all employed. To keep the analysis simple the employers organization does not figure as a player in the game and the money wage increase is assumed to be the instrument at the disposal of the union. The government controls the weighted budget deficit as a fraction of full capacity income.

For two reasons we assume that the trade union is the Stackelberg-leader. *First*, there is an institutional motivation: in collective agreements two parties are involved and once labor contracts are made, they usually are fixed for a certain period. Of course a trade union may try to break open – by strikes and the like – a labor contract. However, we assume that once a labor contract has been agreed upon, the union feels obliged to respect it. Thus the instrument controlled by the union is the nominal wage increase that results from the collective bargaining settlement applying for some fixed period. As the government is autonomous and thus more flexible in its budget decisions (Hersoug, 1985), it is reasonable to consider it to be the follower. *Secondly*, it is plausible to assume that a trade union keeps in mind that government budget decisions are taken in response to the wage increases that have been realized partly by the union's wage claims: often, governments warn the private sector that collective agreements must not let wages get out of hand. It is not likely, on the other hand, that a government should pay much attention to the way a trade union reacts to budgetary policy. For the government, the trade union is just one of the many organizations representing group interests, so the government cannot be expected to take the trade union's response into account.

Assuming the government and the trade union hold different opinions on the working of the economy implies that the models used by these agents in determining their strategy differ. In this study the model differences are expressed in the parameter values of the players models.

We assume that the players use a model for a closed economy with constant technology, where actual unemployment depends on a structural (or natural) component, on the rise in the real wage rate, on autonomous private spending and, finally, on the (weighted) government budget deficit. Monetary policy is used merely to accommodate inflation and so changes in the money supply do not appear in the model. Price-inflation is influenced by both demand-pull and cost-push elements. Prices react with a time-lag of one period to nominal wages.

The parameters in the model used by the government are denoted with superscript G, those in the trade union's model with superscript L. The reduced form model consists of the following three equations:

$$u_t = u^{i*} + \beta^i w_t^r - \alpha^i d_t + \kappa^i \tag{1}$$

$$p_t = \delta^i w_{t-1} - \gamma^i (u - u^{i*}) \tag{2}$$

$$w_t^r \equiv w_t - p_t \tag{3}$$

w_o given

$t = 1, 2 \ldots\ldots$

$i = G, L$

where

u_t = the rate of unemployment
p_t = the rate of inflation
w_t = the rate of change of money wages, control variable of the union
w_t^r = the rate of change of real wages
d_t = the weighted deficit as a fraction of productive capacity, control variable of the government
u^* = the rate of unemployment remaining at full capacity income;

with the suffix t indicating the time index and the superscript i indicating the player's index and with all coefficients assumed to be greater than zero.

We assume that the players agree on the reliability of the data regarding the realized values of unemployment u, price inflation p, and on the realized control variables of the players, the rate of wage inflation w and the weighted deficit as a fraction of capacity, d. The rate of structural unemployment, u^*, is a variable which cannot be unambiguously measured; the same applies to κ, which can be interpreted as the deflationary gap as a fraction of productive capacity. Hence it is possible to disagree upon the true value of these variables as it is over the parameter values. In our view, the equation for unemployment, equation (1) is the one that gives most scope for disagreement. For example, a relatively large value of β^G combined with a relatively low value of κ^G may imply, that the government regards budgetary policy primarily as a tool for counteracting the unemployment effect of wage increases, whereas in the eyes of the trade union fiscal policy is needed to offset the unemployment due to deficient private aggregate demand, implying that the value of κ^L is relatively large. Furthermore, the players can be assumed to have different opinions about the value of u*, which implies that their ideas about the possibility and desirability of reducing actual unemployment differs.

As far as the welfare loss functions are concerned, we have for convenience assumed that these are quadratic, with equal weights applying to the targets that are to be minimized. The government aims equally at minimizing inflation and the difference between actual and structural unemployment. Hence the government's welfare loss function:

$$V^G = \tfrac{1}{2}\sum p_t^2 + \tfrac{1}{2}\sum(u_t - u^{*G})^2 \qquad (4)$$

The trade union is equally interested in reducing the difference between actual and structural unemployment and in reaching a target real wage increase, w*, which in the context of this study is assumed to be exogenously determined. Hence the trade union aims at minimizing

$$V^L = \tfrac{1}{2}\sum(u_t - u^{*L})^2 + \tfrac{1}{2}\sum(w_t^r - w^*)^2 \qquad (5)$$

4. Solutions

4.1 INTRODUCTION

The assumption of trade union leadership implies, that the government takes the change in the money wage rate at time t, w_t, as given when deciding on the optimal value of its instrument in that period, d_t. The government's strategy then consists of finding the deficit as a function of the money wage change that minimizes its welfare loss function, equation (4). In the usual one-model approach the union, with this response in mind, chooses the money wage change that minimizes (5). In the multiple-model analysis, however, the reaction function the trade union has in mind depends on the information available to the trade union about the model used by the government. In the next sections we shall apply the solution concepts introduced in section 2, distinguishing between two optimal pairs of strategies for the trade union and the government, the informed leader solution and the ignorant leader solution. In doing so, we use the method of sequential discrete optimization, which implies that the agents choose their strategies in every period t so as to minimize their welfare loss function in that period, hence the players are assumed to be myopic. This optimization procedure yields solutions which are time-consistent in the

Kydland-Prescott sense, but of course the results are sub-optimal if compared to the open loop, closed loop and closed loop-feedback solutions.

4.2 The Informed Leader Solution

The strategy that, in the government's view, minimizes its welfare loss function results from optimizing (5) subject to the government's model, taking the union's instrument, m, as given. This results in the policy rule of the government:

$$d_t = \varpi_1 w_t - \varpi_2 w_{t-1} + \varpi_3 \qquad (6)$$

where

$$\varpi_1 = \beta^G/\alpha^G$$

$$\varpi_2 = \delta^G(\beta^G + \gamma^G)/[\alpha^G(1 + \gamma^{G2})]$$

$$\varpi_3 = \kappa^G/\alpha^G$$

From equation (6), the government's control varies positively with the deflationary gap as perceived by the government, κ^G, which of course reflects the welfare cost for the government associated with higher unemployment due to deficient aggregate demand. The government's perception of the multiplier, α^G, is in the denominator, because the larger the perceived effect of fiscal policy the less fiscal policy is needed to offset the unemployment effect of wage increases. Both ϖ_1 en ϖ_2 vary positively with β^G reflecting the governments's objective of reducing unemployment and inflation respectively.

In the informed leader solution concept the leader is assumed to know both the model that the follower has in mind and the follower's welfare loss function. Hence in our example the trade union is assumed to know equation (6), the reaction function of the government. Consequently, the trade union determines its optimal strategy by minimizing its welfare function, equation (5), subject to its own model and to the government's response given by (6). This results in the pair of strategies $\hat{w}_t$, $\hat{d}_t$, called the informed leader solution:

$$\hat{w} = \hat{b}_1 w_{t-1} + \hat{b}_2 \qquad (7)$$

$$\hat{d}_t = (\varpi_1\hat{b}_1 - \varpi_2)w_{t-1} + \varpi_1\hat{b}_2 + \varpi_3 \qquad (8)$$

where $\hat{b}_1$ and $\hat{b}_2$ and consequently both $\hat{w}_t$ and $\hat{d}_t$ depend on the parameters of both the model of the trade union and that of the government.

From equation (7) it can be seen, that if both $\hat{b}_1$ and $\hat{b}_2 > 0$, then the optimal strategy for the union implies that nominal wages will keep on rising; if $0 < \hat{b}_1 < 1$, then the rise in nominal wages converges to a constant rate. From (8) it can be seen that if, moreover, $\varpi_1\hat{b}_1 - \varpi_2 > 0$, then it is optimal for the government to permanently run a deficit. As we are in this paper primarily interested in a comparison between the utility levels that come about in the two solution types and as $\hat{b}_1$ and $\hat{b}_2$ are complicated parameters, we shall analyze the implications of this solution, as compared to the ignorant-leader solution derived in the next section, for the players' welfare loss by carrying out numerical simulations (see section 5).

4.3 The Ignorant Leader Solution

The ignorant-leader solution concept introduced in section 2 applies if the Stackelberg leader is not aware of the fact, that the follower uses a different economic model. In that case, as the leader knows the cost function of the follower, the follower's reaction function as perceived by the ignorant leader is derived by optimizing the *follower's cost function* subject to the *leader's* model. In our example, then, the government's policy rule, that is its response at every time t to a given nominal wage change w_t, *as perceived by the trade union*, then is obtained by optimizing the government's welfare loss function (4) subject to the union's model. This results in the government's policy rule as perceived by the trade union, d_t^p :

$$d_t^p = \eta_1 w_t - \eta_2 w_{t-1} + \eta_3 \tag{9}$$

where

$$\eta_1 = \beta^L/\alpha^L$$

$$\eta_2 = \delta^L(\beta^L + \gamma^L)/\alpha^L(1+\gamma^{L2})$$

$$\eta_3 = \kappa^L/\alpha^L$$

Of course, equation (9) is similar to equation (6), because the difference in views between the government and the union are limited to the parameter values of their models. Notice that if union is more keynesian than the government, reflected by $\beta^L/\alpha^L < \beta^G/\alpha^G$, then the expected budgetary policy reaction to a nominal wage claim is larger in the informed leader case than it is in the ignorant leader solution. Hence, in the informed leader case a higher nominal wage claim shall come about than in the ignorant leader case.

With equation (9) in mind the trade union decides on the strategy over time that minimizes its welfare loss function, equation (4). Denoting the optimal strategy for the union with a bar then gives

$$\bar{w}_t = \bar{b}_1 w_{t-1} + \bar{b}_2 \tag{10}$$

where $\bar{b}_1$ and $\bar{b}_2$, and thus $\bar{w}_t$ incorporate merely parameters of the trade union's model, as opposed to $\hat{b}_1$ and $\hat{b}_2$ (see section 4.2), which are influenced by both the government's and the union's model.

Of course the government chooses its optimal strategy with the help of equation (6), that being its true reaction function. This strategy is then given by

$$\bar{d}_t = (\varpi_1\bar{b}_1 - \varpi_2)w_{t-1} + \varpi_1\bar{b}_2 + \varpi_3 \tag{11}$$

where $\bar{d}_t$ depends on parameters of both models, as it is based on both eq.(6) and eq (10). The time-path for the government's optimal strategy in the ignorant-leader solution differs from that in the informed-leader solution because $\bar{b}_1$ and $\bar{b}_2$ differ from $\hat{b}_1$ and $\hat{b}_2$. Moreover, the w_{t-1} that in every period t is given for both players of course differs according to whether the informed-leader solution or the ignorant-leader solution applies.

5. Numerical Simulations

In this section we shall present the results of the numerical simulations that we have carried out in order to analyze the informed leader solution and the ignorant leader solution derived in sections 4.2 and 4.3. We shall, for the reasons given in section 3, concentrate on possible differences between the parameters in the equation for unemployment in the model of the government and the trade union, u^*, α, β and κ.

Of course, the optimal pair of strategies that shall come about in the informed leader situation and the one resulting in the ignorant leader situation can be calculated without making any further assumptions with respect to the true working of the hypothetical economy that the players operate in. However, if we wish to study the *impact* of the strategies chosen by the players *on actual macroeconomic performance* we must introduce a model that is truly describing macroeconomic reality. This also enables us to assess the *realized* values of the welfare loss functions and to compare these with these values as expected by the players.

As far as economic reality is concerned, we assume that one of the players has a correct view on the working of the economy. Of course this assumption is quite arbitrary. It may very well be that both players are wrong and that the economic process can be described by a model with parameters different from those of the trade union as well as those of the government, possibly containing variables that do not appear in the players models. However, in order not to complicate matters further, we avoid introducing a third model. We assume, again quite arbitrarily, that the leader, in our example the trade union, has a model in mind that truly describes the working of the economy.

Thus our simulation results include the pair of strategies that result in the informed leader solution and in the ignorant leader solution, the macroeconomic performance that comes about in these two situations and the expected and realized values of the players' welfare loss functions connected with these two solutions. Of course, the *realized* value of the welfare loss function of the government shall, in both the informed leader solution and the ignorant leader solution, differ from what the government *expects*. The same applies to the value of the trade union's welfare loss function in the ignorant leader case, whereas in the informed leader case because of our assumption that the trade union's model is correctly describing the working of the economy, the realized value corresponds to the trade union's expectations.

In all simulations we have focused on differences in the unemployment equation, eq. (1), assuming that the parameters in the government's equation for price inflation, δ^G and γ^G, are identical with those in the model of the trade union, δ^L and γ^L. Moreover, we have kept the parameters in the trade union's model – which is the model describing economic reality – constant, varying those in the unemployment equation of the government's model.
Thus we have concentrated on the impact of mistakes in the government's view on the relationship between real wage increases and unemployment, β^G, the deflationary gap as a fraction of productive capacity, κ^G, on the rate of structural unemployment, u^{*G} and on the income multiplier of aggregate demand, α^G. As far as the parameters in the model of the trade union, and thus by assumption the model describing economic reality, are concerned, we have assumed: $\alpha = 1.5$; $\beta = 0{,}1$; $\gamma = 0.3$; $u^* = 2$; $\kappa = 8$; $\delta = 0{,}8$. The union is assumed to aim at reaching an increase in the real wage rate of 2%, thus $w^* = 2$.

First we present, as a base-line, the results that would come about if the government's model would be identical to that of the trade union (and, thus, reality). In that case the distinction between the informed leader and the ignorant leader solution is irrelevant and we get a standard one-model result, which is presented in Table 1.

Thus in the conventional one-model situation the union reaches its target real wage change, whereas its unemployment target is not met, creating a small welfare loss for the union. The governments welfare loss is larger, due to the inflation rate of over 5%.

Having derived the one-model Stackelberg game solution as a base-line, we now turn to the simulation results of the multiple-model game. The simulations carried out all are based on the assumption that the government is less keynesian, or more classical, than the labor union. This is reflected in the parameter values of the government's model as compared to those in the union's model. Tables 2–5 (pp. 150–153) show the results of numerical simulations carried out for both the informed-leader case and the ignorant leader case. In Table 2 it is assumed, that the difference between the model of the government and that of the trade union is restricted to the value of α and that the government underestimates this value. Secondly (Table 3) it is assumed, that the government underestimates α and overestimates β. Next, in Table 4, the government overestimates not only the (adverse) effect of rising real wages on unemployment, but also the natural rate of unemployment, u^*. Finally, the government is assumed to overestimate u^* and to underestimate the deflationary gap, κ (Table 5).

TABLE 1. Base-line: identical models.
$\alpha = 1.5 \quad \beta = 0.5 \quad \kappa = 8 \quad \gamma = 0.3 \quad \delta = 0.8 \quad u^* = 2 \quad w^* = 2$

obs	W	D	VL	VG	U	P	WR
2	2.00	5.47	0.00	0.00	2.00	0.00	2.00
3	3.47	5.17	0.10	1.17	2.44	1.47	2.00
4	4.55	4.96	0.39	4.70	2.76	2.55	2.00
5	5.34	4.80	0.89	10.77	3.00	3.34	2.00
6	5.92	4.68	1.58	19.13	3.17	3.92	2.00
7	6.34	4.60	2.43	29.41	3.30	4.34	2.00
8	6.65	4.54	3.40	41.21	3.40	4.65	2.00
9	6.88	4.49	4.48	54.22	3.47	4.88	2.00
10	7.05	4.46	5.63	68.13	3.52	5.05	2.00
11	7.18	4.43	6.83	82.73	3.55	5.18	2.00
12	7.27	4.41	8.08	97.85	3.58	5.27	2.00
13	7.33	4.40	9.36	113.36	3.60	5.33	2.00
14	7.38	4.39	10.66	129.14	3.61	5.38	2.00
15	7.42	4.38	11.98	145.14	3.63	5.42	2.00
16	7.44	4.38	13.32	161.30	3.63	5.44	2.00
17	7.46	4.37	14.66	177.57	3.64	5.46	2.00
18	7.48	4.37	16.01	193.93	3.64	5.48	2.00
19	7.49	4.37	17.37	210.35	3.65	5.49	2.00
20	7.50	4.37	18.73	226.81	3.65	5.50	2.00
21	7.50	4.37	20.09	243.31	3.65	5.50	2.00
22	7.51	4.37	21.45	259.83	3.65	5.51	2.00
23	7.51	4.36	22.82	276.37	3.65	5.51	2.00
24	7.51	4.36	24.19	292.92	3.65	5.51	2.00
25	7.51	4.36	25.55	309.48	3.65	5.51	2.00

Confronting the simulation results in Tables 2–5 with the base-line result shown in Table 1 it can be noticed, first, that the time-path for the union's instrument, w, implies higher wage inflation in the informed-leader cases analyzed here than in the ignorant-leader solution. Of course, in the ignorant leader case the value of w is always equal to the base-line result, as it depends only on the parameters of the union's (and thus the true) model, which is not varied.The time-path for the government's instrument d implies less expansionary policy in the ignorant-leader solution. Price-inflation (a target for the government) in the informed-leader case exceeds that in the ignorant-leader solution, whereas unemployment (an objective for both players) is in the informed-leader case closer to the structural rate of 2% than it is in the ignorant-leader situation. As far as the players' welfare loss is concerned, it can be noticed that in all simulation results presented here for the government it is lower in the ignorant-leader situation. This implies, that it pays for the government to make the union believe that their models are identical. This may be explained as follows. If the union overestimates α as perceived by the government, the union underestimates the weighted deficit that comes about given its nominal wage claim. This may moderate wage demands. Alternatively, if the trade union underestimates β as perceived by the government, this may result in lower wage demands as compared to the informed leader solution because the union, given the government's objective of reducing unemployment, expects less fiscal expansion to compensate for the unemployment due to the rising real wage.[2] However, the government's model is assumed to be wrong and so the welfare loss the government expects to get must differ from the loss that actually is realized. So if we want to know whether there may be an incentive for the follower, the government, to make the union believe that their models are identical we have to focus attention on the welfare loss as expected by the government. From the results presented in Tables 2–5 it can be concluded that in the examples simulated here there actually is an incentive for the follower to hide its model, because in all cases the welfare loss as expected by the government, V^{Ge}, is lower in the ignorant-leader case.[3]

[2] Of course, this incentive for the government to keep the trade union ignorant of their model differences is conditional on the assumption that the government knows the trade union's model, though he believes it to be wrong.

[3] Of course it is also possible to construct numerical examples that do not have this characteristic.

TABLE 2a. Informed-leader solution. $\alpha^g = 0.8$

obs	W	D	VL	VG	U	P	WR	VGE
2	3.56	10.45	28.65	30.97	-5.54	2.26	1.30	0.00
3	6.10	9.46	47.05	60.07	-4.04	4.66	1.44	34.69
4	7.91	8.75	59.51	92.73	-2.97	6.37	1.54	71.00
5	9.20	8.25	68.44	130.41	-2.21	7.59	1.61	111.11
6	10.12	7.89	75.22	172.93	-1.67	8.46	1.66	155.27
7	10.78	7.63	80.65	219.54	-1.28	9.08	1.70	203.01
8	11.24	7.45	85.20	269.40	-1.00	9.52	1.72	253.64
9	11.58	7.32	89.17	321.73	-0.81	9.84	1.74	306.51
10	11.81	7.23	92.76	375.90	-0.67	10.06	1.75	361.07
11	11.98	7.16	96.08	431.45	-0.57	10.22	1.76	416.88
12	12.10	7.12	99.22	487.98	-0.50	10.34	1.77	473.61
13	12.19	7.08	102.24	545.23	-0.45	10.42	1.77	530.99
14	12.25	7.06	105.17	602.99	-0.41	10.48	1.78	588.86
15	12.30	7.04	108.03	661.13	-0.38	10.52	1.78	647.06
16	12.33	7.03	110.85	719.54	-0.37	10.55	1.78	705.52
17	12.35	7.02	113.64	778.13	-0.35	10.57	1.78	764.15
18	12.36	7.01	116.41	836.87	-0.34	10.58	1.78	822.90
19	12.38	7.01	119.16	895.70	-0.34	10.59	1.78	881.75
20	12.38	7.01	121.90	954.59	-0.33	10.60	1.78	940.66
21	12.39	7.00	124.63	1013.54	-0.33	10.61	1.78	999.62
22	12.39	7.00	127.36	1072.52	-0.33	10.61	1.78	1058.61
23	12.40	7.00	130.08	1131.53	-0.32	10.61	1.78	1117.62
24	12.40	7.00	132.80	1190.56	-0.32	10.61	1.78	1176.65
25	12.40	7.00	135.52	1249.60	-0.32	10.62	1.78	1235.69

TABLE 2b. Ignorant leader solution.

obs	W	D	VL	VG	U	P	WR	VGE
2	2.00	10.25	29.82	29.82	-5.40	2.22	-0.22	0.00
3	3.47	9.70	53.54	57.70	-4.56	3.57	-0.10	30.99
4	4.55	9.30	73.23	85.75	-3.94	4.56	-0.01	61.23
5	5.34	9.00	90.22	114.80	-3.49	5.28	0.05	91.82
6	5.92	8.78	105.35	145.05	-3.16	5.82	0.10	123.16
7	6.34	8.62	119.19	176.42	-2.92	6.21	0.13	155.32
8	6.65	8.50	132.12	208.76	-2.74	6.50	0.16	188.23
9	6.88	8.42	144.41	241.87	-2.61	6.71	0.18	221.76
10	7.05	8.36	156.23	275.60	-2.51	6.86	0.19	255.79
11	7.18	8.31	167.72	309.80	-2.44	6.98	0.20	290.20
12	7.27	8.27	178.97	344.35	-2.39	7.06	0.21	324.92
13	7.33	8.25	190.04	379.17	-2.35	7.12	0.21	359.85
14	7.38	8.23	200.98	414.19	-2.33	7.16	0.22	394.96
15	7.42	8.22	211.83	449.36	-2.31	7.20	0.22	430.19
16	7.44	8.21	222.61	484.64	-2.29	7.22	0.22	465.52
17	7.46	8.20	233.34	520.00	-2.28	7.24	0.22	500.91
18	7.48	8.20	244.04	555.42	-2.27	7.25	0.23	536.35
19	7.49	8.19	254.71	590.88	-2.27	7.26	0.23	571.84
20	7.50	8.19	265.35	626.38	-2.26	7.27	0.23	607.34
21	7.50	8.19	275.99	661.90	-2.26	7.27	0.23	642.87
22	7.51	8.19	286.61	697.44	-2.26	7.28	0.23	678.42
23	7.51	8.18	297.23	732.99	-2.25	7.28	0.23	713.97
24	7.51	8.18	307.84	768.55	-2.25	7.28	0.23	749.54
25	7.51	8.18	318.44	804.11	-2.25	7.28	0.23	785.11

TABLE 3a. Informed-leader solution. $\alpha^g = 1$, $\beta^g = 0.2$

obs	W	D	VL	VG	U	P	WR	VGE
2	2.37	8.47	11.15	11.59	-2.61	1.38	0.99	0.00
3	4.20	7.97	18.88	23.62	-1.84	3.05	1.16	13.24
4	5.62	7.58	24.40	38.28	-1.24	4.34	1.29	28.80
5	6.72	7.28	28.46	56.37	-0.78	5.33	1.39	47.56
6	7.57	7.05	31.54	77.95	-0.42	6.10	1.47	69.64
7	8.23	6.87	33.96	102.71	-0.15	6.70	1.53	94.78
8	8.74	6.73	35.92	130.24	0.07	7.16	1.57	122.59
9	9.13	6.62	37.56	160.09	0.23	7.52	1.61	152.66
10	9.44	6.54	38.97	191.85	0.36	7.80	1.64	184.58
11	9.67	6.47	40.21	225.13	0.46	8.01	1.66	218.00
12	9.86	6.42	41.34	259.65	0.54	8.18	1.68	252.61
13	10.00	6.38	42.37	295.13	0.60	8.31	1.69	288.17
14	10.11	6.35	43.34	331.38	0.64	8.41	1.70	324.48
15	10.19	6.33	44.26	368.24	0.68	8.48	1.71	361.38
16	10.26	6.31	45.13	405.56	0.70	8.54	1.72	398.73
17	10.31	6.30	45.99	443.25	0.73	8.59	1.72	436.45
18	10.35	6.29	46.82	481.22	0.74	8.62	1.72	474.44
19	10.38	6.28	47.63	519.42	0.76	8.65	1.73	512.66
20	10.40	6.27	48.43	557.79	0.77	8.67	1.73	551.04
21	10.42	6.27	49.22	596.29	0.77	8.69	1.73	589.55
22	10.43	6.26	50.00	634.90	0.78	8.70	1.73	628.17
23	10.44	6.26	50.77	673.59	0.78	8.71	1.73	666.86
24	10.45	6.26	51.55	712.34	0.79	8.72	1.73	705.62
25	10.46	6.26	52.31	751.14	0.79	8.73	1.73	744.42

TABLE 3b. Ignorant leader solution.

obs	W	D	VL	VG	U	P	WR	VGE
2	2.00	8.40	11.21	11.21	-2.54	1.36	0.64	0.00
3	3.47	7.96	19.53	22.51	-1.87	2.76	0.71	12.39
4	4.55	7.64	26.01	35.39	-1.38	3.79	0.76	26.04
5	5.34	7.40	31.30	50.26	-1.02	4.54	0.79	41.46
6	5.92	7.23	35.79	67.04	-0.76	5.10	0.82	58.62
7	6.34	7.10	39.74	85.46	-0.56	5.50	0.84	77.32
8	6.65	7.00	43.33	105.20	-0.42	5.80	0.86	97.27
9	6.88	6.93	46.65	125.99	-0.32	6.02	0.87	118.20
10	7.05	6.88	49.79	147.59	-0.24	6.18	0.87	139.91
11	7.18	6.85	52.80	169.80	-0.18	6.30	0.88	162.19
12	7.27	6.82	55.72	192.47	-0.14	6.38	0.88	184.92
13	7.33	6.80	58.57	215.48	-0.11	6.45	0.89	207.97
14	7.38	6.79	61.37	238.74	-0.09	6.49	0.89	231.27
15	7.42	6.77	64.13	262.20	-0.07	6.53	0.89	254.74
16	7.44	6.77	66.87	285.79	-0.06	6.55	0.89	278.35
17	7.46	6.76	69.59	309.49	-0.05	6.57	0.89	302.06
18	7.48	6.76	72.29	333.26	-0.05	6.58	0.89	325.84
19	7.49	6.75	74.98	357.09	-0.04	6.59	0.89	349.68
20	7.50	6.75	77.67	380.95	-0.04	6.60	0.89	373.55
21	7.50	6.75	80.35	404.85	-0.03	6.61	0.89	397.45
22	7.51	6.75	83.03	428.77	-0.03	6.61	0.89	421.37
23	7.51	6.75	85.70	452.71	-0.03	6.61	0.89	445.31
24	7.51	6.75	88.38	476.66	-0.03	6.62	0.89	469.26
25	7.51	6.75	91.05	500.62	-0.03	6.62	0.89	493.22

TABLE 4a. Informed-leader solution. $\alpha^g = 1$, $\beta^g = 0.2$, $u^{*g} = 4$

obs	W	D	VL	VG	U	P	WR	VGE
2	2.37	8.47	11.15	22.82	-2.61	1.38	0.99	0.00
3	4.20	7.97	18.88	44.52	-1.84	3.05	1.16	24.47
4	5.62	7.58	24.40	67.68	-1.24	4.34	1.29	49.71
5	6.72	7.28	28.46	93.34	-0.78	5.33	1.39	76.96
6	7.57	7.05	31.54	121.76	-0.42	6.11	1.47	106.60
7	8.23	6.87	33.96	152.82	-0.15	6.70	1.53	138.59
8	8.74	6.73	35.92	186.22	0.07	7.16	1.57	172.71
9	9.13	6.62	37.56	221.61	0.23	7.52	1.61	208.65
10	9.44	6.54	38.97	258.65	0.36	7.80	1.64	246.10
11	9.67	6.47	40.21	297.02	0.46	8.01	1.66	284.80
12	9.86	6.42	41.33	336.46	0.54	8.18	1.68	324.49
13	10.00	6.38	42.37	376.76	0.60	8.31	1.69	364.98
14	10.11	6.35	43.34	417.73	0.64	8.41	1.70	406.10
15	10.19	6.33	44.25	459.23	0.68	8.48	1.71	447.72
16	10.26	6.31	45.13	501.14	0.70	8.54	1.72	489.72
17	10.31	6.30	45.98	543.38	0.73	8.59	1.72	532.03
18	10.35	6.29	46.81	585.87	0.74	8.62	1.72	574.58
19	10.38	6.28	47.63	628.56	0.76	8.65	1.73	617.31
20	10.40	6.27	48.43	671.40	0.77	8.67	1.73	660.18
21	10.42	6.27	49.21	714.36	0.77	8.69	1.73	703.16
22	10.43	6.26	50.00	757.41	0.78	8.70	1.73	746.24
23	10.44	6.26	50.77	800.54	0.78	8.71	1.73	789.37
24	10.45	6.26	51.54	843.72	0.79	8.72	1.73	832.56
25	10.46	6.26	52.31	886.94	0.79	8.73	1.73	875.79

TABLE 4b. Ignorant-leader solution.

obs	W	D	VL	VG	U	P	WR	VGE
2	2.00	8.40	11.21	22.29	-2.54	1.36	0.64	0.00
3	3.47	7.96	19.53	43.32	-1.87	2.76	0.71	23.46
4	4.55	7.64	26.01	64.96	-1.38	3.79	0.76	46.85
5	5.34	7.40	31.30	87.87	-1.02	4.54	0.79	71.02
6	5.92	7.23	35.79	112.16	-0.76	5.10	0.82	96.23
7	6.34	7.10	39.75	137.70	-0.56	5.50	0.84	122.44
8	6.65	7.00	43.33	164.29	-0.42	5.80	0.86	149.51
9	6.88	6.93	46.65	191.71	-0.32	6.02	0.87	177.29
10	7.05	6.88	49.79	219.78	-0.24	6.18	0.87	205.62
11	7.18	6.85	52.80	248.36	-0.18	6.30	0.88	234.39
12	7.27	6.82	55.72	277.31	-0.14	6.38	0.88	263.48
13	7.33	6.80	58.57	306.54	-0.11	6.45	0.89	292.81
14	7.38	6.79	61.37	335.98	-0.09	6.49	0.89	322.33
15	7.42	6.77	64.13	365.58	-0.07	6.53	0.89	351.98
16	7.44	6.77	66.87	395.30	-0.06	6.55	0.89	381.74
17	7.46	6.76	69.59	425.10	-0.05	6.57	0.89	411.57
18	7.48	6.76	72.29	454.96	-0.05	6.58	0.89	441.45
19	7.49	6.75	74.99	484.87	-0.04	6.59	0.89	471.38
20	7.50	6.75	77.67	514.81	-0.04	6.60	0.89	501.33
21	7.50	6.75	80.35	544.78	-0.03	6.61	0.89	531.31
22	7.51	6.75	83.03	574.77	-0.03	6.61	0.89	561.30
23	7.51	6.75	85.70	604.77	-0.03	6.61	0.89	591.31
24	7.51	6.75	88.38	634.78	-0.03	6.62	0.89	621.32
25	7.51	6.75	91.05	664.80	-0.03	6.62	0.89	651.34

TABLE 5a. Informed-leader solution. $\kappa^g = 6$, $u^{*g} = u$

obs	W	D	VL	VG	U	P	WR	VGE
2	2.28	6.23	0.66	5.02	0.85	0.34	1.94	0.00
3	3.92	5.72	0.74	9.75	1.62	1.94	1.98	6.55
4	5.10	5.36	0.75	16.20	2.16	3.09	2.01	14.26
5	5.94	5.10	0.90	24.89	2.56	3.91	2.03	23.83
6	6.54	4.91	1.26	35.70	2.84	4.50	2.04	35.25
7	6.98	4.78	1.80	48.28	3.04	4.92	2.05	48.27
8	7.29	4.68	2.51	62.27	3.19	5.23	2.06	62.57
9	7.51	4.61	3.34	77.34	3.29	5.44	2.07	77.86
10	7.67	4.56	4.27	93.22	3.36	5.60	2.07	93.90
11	7.79	4.53	5.28	109.70	3.42	5.71	2.07	110.50
12	7.87	4.50	6.34	126.62	3.46	5.79	2.08	127.50
13	7.93	4.48	7.45	143.86	3.48	5.85	2.08	144.80
14	7.97	4.47	8.58	161.34	3.50	5.89	2.08	162.31
15	8.00	4.46	9.73	178.98	3.52	5.92	2.08	179.98
16	8.02	4.45	10.90	196.74	3.53	5.94	2.08	197.77
17	8.04	4.45	12.09	214.59	3.53	5.96	2.08	215.63
18	8.05	4.45	13.27	232.50	3.54	5.97	2.08	233.55
19	8.06	4.44	14.47	250.46	3.54	5.98	2.08	251.52
20	8.06	4.44	15.67	268.45	3.55	5.98	2.08	269.51
21	8.07	4.44	16.87	286.46	3.55	5.99	2.08	287.53
22	8.07	4.44	18.07	304.49	3.55	5.99	2.08	305.56
23	8.07	4.44	19.28	322.53	3.55	5.99	2.08	323.60
24	8.07	4.44	20.49	340.58	3.55	5.99	2.08	341.65
25	8.07	4.44	21.70	358.63	3.55	5.99	2.08	359.71

TABLE 5b. Ignorant-leader solution.

obs	W	D	VL	VG	U	P	WR	VGE
2	2.00	4.13	2.32	0.19	4.06	-0.62	2.62	0.00
3	3.47	3.84	5.64	0.68	4.50	0.85	2.62	1.37
4	4.55	3.62	9.82	2.88	4.83	1.93	2.62	4.21
5	5.34	3.47	14.70	7.13	5.06	2.72	2.62	8.94
6	5.92	3.35	20.13	13.34	5.24	3.30	2.62	15.49
7	6.34	3.26	25.98	21.20	5.36	3.72	2.62	23.61
8	6.65	3.20	32.15	30.41	5.46	4.04	2.62	33.01
9	6.88	3.16	38.57	40.67	5.53	4.27	2.62	43.41
10	7.05	3.12	45.16	51.75	5.58	4.43	2.62	54.59
11	7.18	3.10	51.88	63.44	5.61	4.56	2.62	66.35
12	7.27	3.08	58.70	75.59	5.64	4.65	2.62	78.56
13	7.33	3.07	65.60	88.09	5.66	4.72	2.62	91.09
14	7.38	3.06	72.55	100.84	5.68	4.76	2.62	103.88
15	7.42	3.05	79.54	113.78	5.69	4.80	2.62	116.84
16	7.44	3.04	86.56	126.86	5.70	4.83	2.62	129.94
17	7.46	3.04	93.60	140.05	5.70	4.85	2.62	143.13
18	7.48	3.04	100.65	153.31	5.71	4.86	2.62	156.41
19	7.49	3.04	107.72	166.63	5.71	4.87	2.62	169.73
20	7.50	3.03	114.79	179.99	5.71	4.88	2.62	183.09
21	7.50	3.03	121.88	193.38	5.71	4.88	2.62	196.48
22	7.51	3.03	128.96	206.79	5.71	4.89	2.62	209.90
23	7.51	3.03	136.05	220.22	5.71	4.89	2.62	223.33
24	7.51	3.03	143.15	233.66	5.72	4.89	2.62	236.77
25	7.51	3.03	150.24	247.10	5.72	4.89	2.62	250.22

6. Concluding Remarks

In this paper a multiple-model Stackelberg game between a trade union and a government was studied. It was argued that macroeconomic applications of game theory should pay attention to the possibility of different views, and thus models, of the players. Moreover it was contended, that in a multiple-model game the information of the leader about the model used by the follower affects the strategies chosen by both players.

Two adaptations of the well-known Stackelberg solution concept were introduced, each based on a specific assumption regarding the information disposable to the leader. Assuming the leaders model to be true, numerical simulations were carried out directed at comparing the informed-leader and the ignorant-leader solutions. Assuming the differences between the view to be restricted to the parameters in the unemployment equation, simulations were carried out focusing on these parameters separately.

Focusing on the welfare loss of the players it was concluded, that there may be an incentive for the follower to cheat, making the leader believe that the models they use are identical. We believe this to be an important issue in the context of the game-theoretic analysis of macroeconomic performance.

Of course, the results presented here are conditional on the model and on the parameter values. Moreover, the method of sequential discrete optimization applied here creates suboptimal results and it is assumed that the players, when confronted with economic results that do not correspond to their model prediction, do not alter their model and strategy. Further research should be directed at eliminating these restrictive assumptions.

References

Calmfors, L., "Employment Policies, Wage Formation and Trade Unions in a Small Open Economy", *The Scandinavian Journal of Economics* 84 (2) 345-372, 1982

Driffill, E.J., "Macroeconomic Stabilization Policy and Trade Union Behavior as a Repeated Game", *Scandinavian Journal of Economics* 87 (2) 1985, 300-326

Frankel, J. A., "Obstacles to International Macroeconomic Policy Coordination", *Princeton Studied in International Finance* no. 64, December 1988

Frankel, J.A., and K. Rockett, "International Macroeconomic Policy Coordination When Policy Makers Do Not Agree on the True Model, *American Economic Review* vol.78 no. 3, 318-340

Ghosh, A., "International Policy Coordination in an Uncertain World", *Economics Letters* 21 (1986) 271-276

Ghosh, A. and P.R. Masson, "International Policy Coordination in a world with Model Uncertainty", *IMF Staff Papers*, 35 (June 1988)

Gylfason, T., and A. Lindbeck, "Endogenous Unions and Governments, A Game-Theoretic Approach", *European Economic Review* 30 (1986), 5-26

Holtham, G., and A. Hughes Hallett, "International Policy Cooperation and Model Uncertainty", in Bryant and Portes (eds), *Global Macroeconomics, Policy Conflict and Cooperation*, Macmillan, 1987

Klerk, R.A. de, and H.M. Prast, "A Note on Rival Views: the Interaction of Fiscal Policy and Wage Determination", *European Journal of Political Economy* 7 (1991), 159-167.

Persson, T., and L. Svensson, "Misperceptions, Rigidity and Welfare", *European Economic Review* 25 (1984) 387-399

Pohjola, M., "Applications of Dynamic Game Theory to Macroeconomics", in T. Basar (ed), *Dynamic Games and Applications in Economics*, Springer, Berlin 1986

EXCHANGE RATE UNCERTAINTY IN IMPERFECT MARKETS: A SIMULATION APPROACH

HANS M. AMMAN AND LIDWIN M.T. VAN VELDEN

Department of Macroeconomics, University of Amsterdam, Jodenbreestraat 23, room 3359, 1011 NH Amsterdam-NL, Email economics@hasara11.bitnet

Abstract. In this paper a model is presented to analyze the influence of exchange rate uncertainty under imperfect market conditions. We consider an oligopolistic market with two types of producers: home producers and foreign producers, both supplying the home market. The exchange rate influences the profits of the foreign producers, and therefore, determines foreign market supply. Through the common demand equation and the game-theoretical setting, this foreign supply affects the home producers.

By assuming profit maximization and allowing the exchange rate to follow an Ito diffusion process, we end up with a stochastic control problem. The closed loop solution of the control problem can be derived by means of the Hamilton-Jacobian-Bellman equation of stochastic control, resulting in one or more partial differential equations (PDEs). As most PDEs are difficult to solve analytically, we apply a simulation approach to investigate the characteristics of our model.

1. Introduction

In recent years there has been a renewed interest in economic literature for imperfect markets in international trade. Important papers in this respect are, for instance, Dixit (1984, 1989), Ethier (1987) and Eaton and Grossman (1986). A somewhat alternative approach can be found in Helpman (1981), who extends the Heckscher-Ohlin model for the case of imperfect markets. Fershtmann and Kamien (1987) developed a duopoly model under the assumption of sticky prices.

The role of exchange rate uncertainty in an imperfect market, however, has been scarcely analyzed. Dixit (1989) has recently made an attempt by presenting a model to explain import-penetration under exchange rate uncertainty. Some empirical work has been done by Gagnon (1989). In this paper we present, as Venables (1990) calls it, a segmented market-approach model to investigate the effects of exchange rate uncertainty in an oligopolistic market[1]. Rather than making special assumptions in order to solve our model analytically, we try to solve the general model using numerical analysis.

The practical relevance of this analysis lies in the fact that in the last couple of years we could observe that, for instance, the value of the US dollar affected world trade flows. The unprecedented high level of the US dollar in the first half of the 1980's made it possible for Japanese and European producers to obtain a substantial market share in the US, which had a major impact on the US trade balance.

[1] An earlier version of this model was presented in Amman (1988).

H.M. Amman et al. (eds), Computational Economics and Econometrics, 157–163.

In Section 2 we present a model for an imperfect market and analyze the effect of a change in the exchange rate on market conditions. Section 3 contains a number of numerical experiments in order to investigate the properties of our model. The conclusions are stated in Section 4.

2. The model

Suppose we have a market with m home producers and n foreign producers, both types of producers operating in the market of the home producers. Under Cournot competition the product supplied by both producers is homogeneous, and so, both producers are confronted with the same price demand function. If we assume that the producers seek profit maximization, the objective function of the home producers can be expressed as

$$V_i = \max \; E_t \int_t^T \pi_i \, \exp(-\lambda_i s) \, ds \qquad i = 1, \ldots, m \tag{1}$$

and accordingly for the foreign producers

$$V_j = \max \; E_t \int_t^T \pi_j \, \exp(-\lambda_j s) \, ds \qquad j = 1, \ldots, n \tag{2}$$

where E_t denotes the expectations operator, λ the discount rate, t the time variable and T the end of the planning period. The profit functions of the home producers, π_i, and the foreign producers, π_j, are specified as

$$\pi_i = \; p \, q_i - c_i \qquad \forall i \tag{3}$$

$$\pi_j = e p \, q_j - c_j \qquad \forall j \tag{4}$$

where

$\pi_{i/j}$ = profit function of home/foreign producers

p = the price on the home market

$q_{i/j}$ = quantity supplied by the home/foreign producers

$c_{i/j}$ = cost function of the home/foreign producers

e = exchange rate denoted as the amount of foreign currency per unit of home currency

The profits of the foreign producers are being affected by the level of the exchange rate. All foreign producers operate with the same exchange rate e.

A common approach is to assume that the following continuous and differentiable functional relationships hold

$$p = p(\ell^T q) \qquad p > 0 \quad p' < 0 \qquad q \geq 0 \tag{5}$$

$$c_i = c_i(q_i) \qquad c_i \geq 0 \quad c_i' > 0 \qquad \forall i \tag{6}$$

$$c_j = c_j(q_j) \qquad c_j \geq 0 \quad c_j' > 0 \qquad \forall j \tag{7}$$

where $q = [q_{i=1} \,..\, q_{i=m}\, q_{j=1} \,..\, q_{j=n}]^T$, $\ell^T = [1\ 1\ ..\ 1]$ an $(n+m)$-element vector and T is used as a transposed sign. Evidently, $q \geq 0$ implies that $\forall i\ q_i \geq 0$ and $\forall j\ q_j \geq 0$. The cost functions of the foreign producers, c_j, are expressed in their own (foreign) currency. Following Dixit (1989) and Svensson (1989), we will assume that we have a freely floating exchange rate which can be described by an Ito diffusion process (cf. Øksendal(1985)).

$$de = \delta(e,t)\, dt + \sigma(e,t)\, dw \qquad e > 0 \tag{8}$$

where w follows a Wiener process with the standard properties $E(dw) = 0$ and $E(dw^2) = dt$. Through the exchange rate equation (8) a source of uncertainty is introduced into the model.

Given equations (1), (2) and (8) the optimal output levels of the home producer and foreign producers can be obtained with the help of the Hamilton-Jacobian-Bellman equation (HJB) of stochastic control (cf. Malliaris and Brock(1982) or Kamien and Schwartz (1981)). In our case the HJB-equations are defined for the home producers as

$$-\frac{\partial V_i}{\partial t} = \max_{q_i} \left[\exp(-\lambda_i t)\, \pi_i(q) + \delta(e,t)\, \frac{\partial V_i}{\partial e} + \tfrac{1}{2}\, \sigma(e,t)^2\, \frac{\partial^2 V_i}{\partial e^2}\right] \quad \forall i \tag{9}$$

and for the foreign producers as

$$-\frac{\partial V_j}{\partial t} = \max_{q_j} \left[\exp(-\lambda_j t)\, \pi_j(q) + \delta(e,t)\, \frac{\partial V_j}{\partial e} + \tfrac{1}{2}\, \sigma(e,t)^2\, \frac{\partial^2 V_j}{\partial e^2}\right] \quad \forall j \tag{10}$$

The equations of (9) and (10) describe a system of $m + n$ partial-differential equations (PDEs). These PDEs are interconnected through the profit term which incorporates the price equation.

In a Cournot-Nash situation each home producer and foreign producer will determine his or her optimal output level applying equations (9) and (10). The first order conditions for profit maximization of each producer can be derived from (3) and (4), that is

$$\partial \pi_i / \partial q_i = p'\, q_i + p - c_i' = 0 \qquad \forall i \tag{11}$$

$$\partial \pi_j / \partial q_j = ep'\, q_j + ep - c_j' = 0 \quad \forall j \tag{12}$$

resulting in the system

$$q = \begin{bmatrix} \phi_1 I_m & 0 \\ 0 & \phi_2 I_n \end{bmatrix} \begin{bmatrix} \psi_i \\ \psi_j \end{bmatrix} \tag{13}$$

or more compactly

$$q = A(q)\,\Psi(q) \tag{13'}$$

where $\psi_i = [p - c'_{i=1} \,..\, p - c'_{i=m}]^T$, $\psi_j = [ep - c'_{j=1} \,..\, ep - c'_{j=n}]^T$, $\phi_1 = -1/p'$, $\phi_2 = -1/ep'$ and I being the identity matrix. Equation (13) states that the marginal revenues should be equal to the marginal costs for each producer. Due to the assumptions with respect to p, e, c_i and c_j equation (13') has a solution q^* for q. Unfortunately, there is no guarantee that $q^* \geq 0$, because of the possibility that for some k, $q_k < 0$. Let's assume, however, that $\forall i\ p \geq c'_i$ and $\forall j\ ep \geq c'_j$, then certainly $q^* \geq 0$. Note that q^* depends on the level of the exchange rate, so $q^* = q^*(e)$. If we insert this optimal solution in the profit functions we get

$$-\frac{\partial V_i}{\partial t} = \left[\exp(-\lambda_i t)\,\pi_i(q^*) + \delta(e,t)\,\frac{\partial V_i}{\partial e} + \tfrac{1}{2}\,\sigma(e,t)^2\,\frac{\partial^2 V_i}{\partial e^2}\right] \quad \forall i \tag{9'}$$

$$-\frac{\partial V_j}{\partial t} = \left[\exp(-\lambda_j t)\,\pi_j(q^*) + \delta(e,t)\,\frac{\partial V_j}{\partial e} + \tfrac{1}{2}\,\sigma(e,t)^2\,\frac{\partial^2 V_j}{\partial e^2}\right] \quad \forall j \tag{10'}$$

As is clear from equations (9') and (10'), the objective V is a function of the exchange rate and time and therefore we have to determine $V(e,t)$ analytically or numerically. Apart from some special cases, no general solution exists for (9) and (10), which makes it necessary to apply numerical analysis to determine $V(e,t)$. This will be done in the next chapter.

A Numerical Experiment

The PDEs denoted in equations (9') and (10') belong to the class of parabolic partial-differential equations with variable coefficients. The famous heat equation in physics is a special case of this type of PDE. There are several methods for solving this type of PDE. One of the most popular ways is by applying the Crank-Nicolson method. This implicit differencing scheme is an unconditionally stable method (cf. Ames(1977)) to solve parabolic PDEs of the form (9) and (10)[2]. We shall use this method to examine the next example.

[2] Nowadays, most software libraries, like IMSL and NAG, contain procedures to solve different kinds of PDEs.

A Numerical Example

Assume we have a duopoly situation in which there is one home producer and one foreign producer both seeking profit maximization. In this case the objective functions of both producers for a time horizon of 100 might look like

$$V_{i=1} = \max E_0 \int_0^{100} \pi_{i=1} \exp(-0.05s)\, ds$$

$$V_{j=1} = \max E_0 \int_0^{100} \pi_{j=1} \exp(-0.05s)\, ds\,.$$

Following Fershtman and Kamien (1987), the possible price equation and costs equations are

$$p = -(q_{i=1} + q_{j=1}) + 10$$

$$c_{i=1} = q_{i=1}$$

$$c_{j=1} = q_{j=1}$$

and the exchange rate equation (without drift, $\delta = 0$)

$$de = \sigma\, dw$$

with

$$\sigma \in \{0.05,\ 0.50,\ 0.75\}$$

and

$$E(e) \in \{0.50,\ 1.00,\ 1.50,\ 2.00,\ 2.50\}\,.$$

By changing the value of the parameter σ we change the level of exchange rate uncertainty. The higher the value of σ the higher the variance of the exchange rate.

We have run a number of experiments with several initial states of the exchange rate. Tables I, II, and III do show the *expected* outcomes of the objective functions, price level and production levels given the various degrees of exchange rate uncertainty. It is clear that a higher level of the exchange rate will lead to a higher market share for the foreign producer, independently of the level of uncertainty. Furthermore, the expected price level and the expected supply levels are not affected by the exchange rate uncertainty. However, the expected profits of *both* producers decline when the exchange rate uncertainty increases. Although both producers are affected by the rise in exchange rate uncertainty, the decline in profits for the foreign producer becomes larger when the exchange rate increases.

TABLE I
Expected values for $\sigma = 0.05$

e	0.50	1.00	1.50	2.00	2.50
p	4.33	4.00	3.89	3.83	3.80
$q_{i=1}$	3.33	3.00	2.89	2.83	2.80
$q_{j=1}$	2.33	3.00	3.22	3.33	3.40
$V_{i=1}$	165.50	184.27	169.19	160.99	158.39
$V_{j=1}$	75.30	183.42	309.09	427.10	486.27

TABLE II
Expected values for $\sigma = 0.25$

e	0.50	1.00	1.50	2.00	2.50
p	4.33	4.00	3.89	3.83	3.80
$q_{i=1}$	3.33	3.00	2.89	2.83	2.80
$q_{j=1}$	2.33	3.00	3.22	3.33	3.40
$V_{i=1}$	26.67	50.14	68.50	80.27	84.55
$V_{j=1}$	50.81	97.05	134.31	158.82	167.92

TABLE III
Expected values for $\sigma = 0.50$

e	0.50	1.00	1.50	2.00	2.50
p	4.33	4.00	3.89	3.83	3.80
$q_{i=1}$	3.33	3.00	2.89	2.83	2.80
$q_{j=1}$	2.33	3.00	3.22	3.33	3.40
$V_{i=1}$	2.58	4.80	6.52	7.62	8.02
$V_{j=1}$	4.81	9.21	12.77	15.13	16.01

3. Conclusions

In this paper we have presented a model of an oligopolistic market structure distinguishing home producers and foreign producers. Instead of making rigerous assumptions in order to solve the model analytically, we applied an differencing scheme to solve the HJB equation numerically. As one might expect, the model seems to indicate that positive changes in the exchange rate benefit the foreign producer. Additionally, exchange rate uncertainty does not only influence the foreign producer, but also the home producer. This means that a higher degree of exchange rate uncertainty has a negative influence on the home producer's profits.

References

Ames, W.F., 1977, *Numerical methods for partial differential equations*, New York: Academic Press.

Amman, H.M., 1988, 'The role of the exchange rate in an oligopolistic market', Research Memorandum 8819, Faculty of Economics, University of Amsterdam.

Dixit, A., 1984, 'International trade policies for oligopolistic industries', *Economic Journal* (supplement), 1–16.

Dixit, A., 1989, 'Hysteresis, import penetration and exchange rate pass-through', *Quarterly Journal of Economics* **54**, 205–228.

Eaton, J. and G.M. Grossman, 1986, 'Optimal trade and industrial policy under oligopoly', *Quarterly Journal of Economics* **51**, 383–406.

Ethier, W.J., 1987, 'The theory of International Trade'. In L.H. Officer, *International Economics*, Boston: Kluwer Academic Publishers, 1–57.

Fershtman, C. and M.I. Kamien, 1987, 'Dynamic duopolistic competition with sticky prices', *Econometrica* **55**, 1151–1164.

Gagnon, J.E., 1989, 'Exchange rate variability and the level of international trade', International Finance discussion paper 369, Board of Governors of the Federal Reserve System, Washington, USA.

Helpman, E., 1981, 'International trade in the presence of product differentiation, economies of scale and monopolistic competition', *Journal of International Economics* **11**, 305–340.

Kamien, M.I. and N.L. Schwartz, 1981, *Dynamic optimization*, Amsterdam: North-Holland.

Malliaris, A.G. and W.A. Brock, 1982, *Stochastic methods in economics and finance*, Amsterdam: North-Holland.

Øksendal, B., 1985, *Stochastic differential equations*, Berlin: Springer-Verlag.

Svensson, L.E.O., 1989, 'Target zones and interest rate variability.' Seminar paper 457, Institute for International Economic Studies, Stockholm, Sweden.

Venables, A.J., 1990, 'The economic integration of oligopolistic markets', *European Economic Review* **34**, 753–773.

SUBJECT INDEX

Acceleration technique 13–15
Algorithm, equilibration 67–106
  global optimization 21–32
  multivariate optimization 23–32
  Paige's 46–54
  splitting equilibration 65–106
  USA 107–120
Analysis, game-theoretic 139–154
Annealing, simulated 19–32
Antithetic Variate (AV) 13–15
Approach, multiple-model 138–155
ARMA model 33–42
  parameter 32–42

Benchmark, Linpack 30
Bernoulli drawing 11–13
Bivariate random variable 10–13
Brute force simulation 9–13

Cheating 141–154
Common Random Number (CRN) 15–18
Competition, Cournot 158–163
Computational cost 107–120
  economics 1
  performance 67–106
Computing, high performance 23–32
Concept, solution 139–155
Condition, rank 58–62
Conditional 4–18
Constrained matrix problem 65–106
Contraction, inside 127–136
Control, stochastic 157–163
  Variate (CV) 14 15
Convergence, relative convergence speed and 115–120
  verification 80–106
Correction mechanism, error 3
Cost, computational 107–120
Cournot competition 158–163

Decomposition, QR 45–54
Deficiency, rank 43–54
Degree of effective identification 55–62
Determination, wage 137–155
Diagnostic measure 57–62
Diagonal constrained matrix problem 72–106
  dominance 107–120
  model 70–106
Dominance, diagonal 107–120
Drawing, Bernoulli 11–13
Dynamic disequilibrium model 10–18
  Latent Variable (DLV) 3–18
  structure 7–13
  simplifying 4–18

Econometric model 122–136
Economic performance 139–155
Economics, computational 1
  regional 69–106
Equation, estimator generation 6–18
  Hamilton-Jacobian-Bellman 157–163
  Lyapunov 33–42
  parabolic partial-differential 160–163
  partial differential 157–163
Equilibration algorithm 67–106
  exact 66–106
Error correction mechanism 3
Estimation, maximum likelihood 34–42
Estimator, generation equation 6–18
  Least Generalized-Variance Ratio 59–62
  least-squares 55–62
  maximum likelihood 4–18
  Swamy and Tinsley's 43–54
Exact equilibration 66–106
Exchange rate, freely floating 159–163
Expectation, unconditional 35–42
Experiment, numerical 158–163
  parallel 103–106
  serial 93–106

Faculty Fellowship Award 104
Fast Gauss-Seidel (FGS) method 109–120
Filter, Kalman 33–42, 46–54
Fiscal policy 137–155
FORTRAN, Parallel 99–106
Freely floating exchange rate 159–163
Function, importance 4–18
  likelihood 4–18, 21–32, 33–42, 51–54
  objective 162, 163
  simulated likelihood 7–13
  test 24–32
  transportation cost 98–106

Game, Stackelberg 137–155
  -theoretic analysis 139–154
Gauss-Seidel iteration 107–120, 122–136
Global optimization algorithm 21–32

Hamilton-Jacobian-Bellman equation 157–163
Hankel-plus-Toeplitz structure 39–42
High performance computing 23–32

Identifiable 55–62
Identification, degree of effective 55–62
Ignorant leader solution 140–154
Importance function 4–18
Information 138–155
Informed leader solution 140–154
Inside contraction 127–136
Integration, Monte Carlo 4 8–13
Iteration, Gauss-Seidel 107–120
Itself, likelihood function 4–18
I/O matrix 97–106

Jacobi, Gauss-Seidel and Successive Overrelaxation (SOR) method 108–120
Jacobi overrelaxation (JOR) method 109–120

Kaji, I. 72
Kalman filter 33–42, 46–54
Kronecker product 45–54

Large Optimizing System (LOS) 121–136
Large-scale constrained matrix problem 65–106
  nonlinear macroeconomic model 107–120
Least Generalized-Variance Ratio Estimator (LGVRE) 59–62
Least-squares estimator 55–62
Levenberg-Marquardt method 48–54
Likelihood function 4–18, 21–32, 33–42, 51–54
  itself 4–18
Limited Dependent Variable (LDV) 6–18
Line search 130–136
Linpack benchmark 30
London Business School model 113–120
Lyapunov equation 33–42

Macroeconomic performance 146–154
Market, oligopolistic 157–163
Matrix, constrained 66–106
  I/O 97–106
  social accounting 70–106
Maximum likelihood estimation 4–18, 34–42
Measure, diagnostic 57–62
Method, Fast Gauss-Seidel (FGS) 109–120
  Gauss-Seidel 122–136
  Jacobi, Gauss-Seidel and Successive Overrelaxation (SOR) 108–120
  Jacobi Overrelaxation (JOR) 109–120
  Levenberg-Marquardt 48–54
  maximum likelihood 34–42
  Nelder-Mead 121–136
  no-derivative 121–136
  Simulated Moment (MSM) 4–18
  Turn-the-worst 127–136
Migration table 96–106
Minimization procedure 123–136
Model, diagonal 70–106
  dynamic disequilibrium 10–18
  econometric 122–136
  large-scale nonlinear macroeconomic 107–120
  London Business School 113–120
  NIESR 113–120
  nonlinear econometric 121–136
  segmented market-approach 157–163
  stochastic coefficients 44–54
  structure 107–120
  UK Treasury 113–120
Modelling, symbolic 121–136
Moment (MSM), Method of Simulated 4–18
Monte Carlo integration 4 8–13
Morrison and Marquardt's result 52–54
Multiple-model approach 138–155
Multivariate optimization algorithm 23–32

Nelder-Mead method 121–136
  versus Turn 131–136
NIESR model 113–120
No-derivative method 121–136
Nonlinear econometric model 121–136
  econometric system 108–120
  system 107–120
Nonstationary model solution technique 107–120
Number (CRN), Common Random 15–18
Numerical experiment 158–163

Objective function 162
Observable scalar random variable 44–54
  surrogate for latent variable 4–18
Oligopolistic market 157–163

Optimization procedure 122–136

Paige's algorithm 46–54
Parabolic partial-differential equation 160–163
Parallel experiment 103–106
  FORTRAN (PF) 99–106
Parallelizable 65–106
Parameter, ARMA 32–42
Partial differential equation (PDE) 157–163
Performance, computational 67–106
  economic 139–155
  macroeconomic 146–154
Policy, fiscal 137–155
Problem, constrained matrix 65–106
  diagonal constrained matrix 72–106
  large-scale constrained matrix 65–106
  spatial price equilibrium 67–106
Procedure, minimization 123–136
  optimization 122–136
Process, vector stochastic 43–54
Product, Kronecker 45–54

QR decomposition 45–54

Rank condition 58–62
  deficiency 43–54
Regional economics 69–106
Repeated simplex setting 125–136
Result, Morrison and Marquardt's 52–54

Schnabel 20
Schwartz 21
Search, line 130–136
Segmented market-approach model 157–163
Serial experiment 93–106
Setting, repeated simplex 125–136
Simplifying the dynamic structure 4–18
Simulated annealing 19–32
  likelihood function 7–13
Simulation 4–18, 107–120, 121–136, 146–154
  brute force 9–13
Social accounting matrix (SAM) 70–106
Solution concept 139–155
  ignorant leader 140–154
  informed leader 140–154
Spatial price equilibrium problem 67–106
Speed, convergence and relative convergence 115–120
Splitting equilibration algorithm (SEA) 65–106
Stackelberg game 137–155
Stochastic coefficients model 44–54
  control 157–163
Structure, dynamic 7–13
  Hankel-plus-Toeplitz 39–42
  model 107–120
Stuart, A. 28
Swamy and Tinsley's estimator 43–54
Symbolic modelling 121–136
System, Large Optimizing 121–136
  nonlinear 107–120

Table, migration 96–106
Technique, acceleration 13–15
  nonstationary model solution 107–120
Test function 24–32
Transportation cost function 98–106
Turn-the-worst method 127–136
Turns, Nelder-Mead versus 131–136

UK Treasury model 113–120
Unconditional expectation 35–42
USA algorithm 107–120

Variable, bivariate random 10–13
  Dynamic Latent 3–18
  Limited Dependent 6–18
  observable scalar random 44–54
Variate, Antithetic 13–15
  Control 14 15
Vector stochastic process 43–54
Verification, convergence 80–106

Wage determination 137–155

AUTHORS' INDEX

Aigner, D.J. 3, 17
Aluffi-Pentini, F. 21, 28
Ames, W.F. 161, 163
Amman, H.M. 157–163
Andrews, D.W.K. 6, 17
Ansley, C.F. 33, 37, 38, 41

Bacharach, M. 65, 66, 105
Bachem, A. 67, 72, 101, 105
Barabás, G. 121–136
Bell, G. 28, 31
Belsley, D.A. 43, 55–62
Bertsekas, D.P. 86, 105
Blair, P.D. 69, 106
Bohachevsky, I.O. 21, 28
Brock, W.A. 159, 163
Buchanan, I. 69, 70, 105
Byron, R.P. 71, 97, 105

Calmfors, L. 137, 154
Carnevali, P. 21, 28
Conway, R.K. 44, 54
Corana, A. 20, 23, 27, 28
Cottle, R.W. 72, 86, 105
Cramer, J.S. 19, 28
Crown, W.H. 67, 106
Cvetkovic, L. 108, 110, 112, 113, 119

Dafermos, S. 90, 99, 105
Danielsson, J. 3
Davidson, R. 14, 17
De Melo, J. 97, 105
Deming, W.E. 67, 72, 105
Dempster, A.P. 43, 53
Dennis, J.E. 20, 28
Dervis, K. 97, 105
Derwent, D. 21, 28
Dixit, A. 157, 159, 163
Dongarra, J.J. 28 30
Drexl, A. 21, 28
Driffill, E.J. 137, 154
Duffie, D. 6, 17
Duvall, S.G. 86, 105

Eaton, J. 157, 163
Engle, R.F. 16, 17
Enke, S. 67, 105
Ethier, W.J. 157, 163
Eydeland, A. 65–106

Fair, R.C. 113, 119
Ferrier, G. 19–32
Fershtman, C. 157, 161, 163
Finch, S.J. 19, 28
Fisher, P.G. 107–120
Frankel, J.A. 138, 139, 154
Friedlander, D. 72, 105

Gagnon, J.E. 157, 163
Gardner, G. 33, 36, 41
Geman, D. 6, 17
Geman, S. 6, 17
Geweke, J. 3, 4, 13, 14, 17
Ghosh, A. 138, 154
Goffe, W.L. 19–32
Goldberg, D.E. 21, 29
Goldberger, A.S. 59, 62
Granger, C.W.J. 5, 9, 16, 17
Grossman, G.M. 157, 163
Gylfason, T. 137, 138, 154

Hadjidimos, A. 109, 119
Hajivassiliou, V.A. 6, 17
Hall, A.D. 33, 37, 38, 42
Hallahan, C. 43–54
Hansen, L.P. 6, 17
Hanson, K.A. 97, 105
Hanson, R.J. 48, 49, 53
Harrigan, F. 69, 70, 105
Havenner, H. 47, 53
Heckman, J.J. 3, 4, 17
Helpman, E. 157, 163
Hendry, D.F. 3–18
Herceg, D. 108, 110, 112, 113, 119
Hersoug 137, 141
Holtham, G. 138, 154
Horn 137
Hughes Hallett, A. 107–120, 138, 154

I-Lok Chang 43–54

Johnston, J. 57 62

Jones, R.H. 33, 41
Judge, G.G. 20, 24, 28, 67, 69, 105

Kaji, I. 86, 106
Kamien, M.I. 157, 159, 161, 163
Kendall, M. 21, 28
Khilnani, A. 108, 111–115, 120
Kim, D.-S. 67, 71, 82, 101–103, 105
King, B.B. 97, 105
Kirkpatrick, S. 21 28
Klincewicz, J. 72, 106
Kohn, R. 33, 38, 41, 42
Korte, B. 67, 72, 101, 105
Kourouklis, S. 46, 47, 53

Laffont, J.J. 4, 18
Lawson, C.L. 48, 49, 53
LeBlanc, M.R. 44, 54
Lehmann, E.L. 50, 51, 53
Lerman, S. 6, 18
Lindbeck, A. 137, 138, 154

MacKinnon, J.G. 14, 17
Maddala, G.S. 4, 17
Malinvaud, E. 55 62
Malliaris, A.G. 159, 163
Manski, C. 6
Marquardt, D.W. 52, 53
Masson, P.R. 138, 154
McCurdy, T.F. 4, 17
McFadden, D.L. 4–6, 17
Mélard, G. 33, 41
Miller, R.E. 69, 106
Mittnick, S. 33–42
Mohr, M. 67, 106
Montfort A. 4, 18
Morf, M. 33, 42
Morrison, D.D. 52, 53
Mosteller, F. 69, 106

Nagurney, A. 65–106
Newbold 37
Nicholls, D. 33, 37, 38, 42

Oeksendal, B. 159, 163
Ohuchi, A. 72, 86, 106
Ostrowski, A. 110, 120

Paige, C.C. 46, 47, 51, 53
Pakes, A. 4–6, 18
Pearlman, J.G. 33, 42
Pohjola, M. 138, 155
Polenske, K.E. 67, 69, 104, 105
Pollard, D. 4–6, 18
Prast, H. 137–155
Press, W.H. 20, 21, 27, 28
Priestley, M.B. 46, 53
Pronzato, L. 21, 28
Pyatt, G. 70, 106

Quandt, R.E. 3, 4, 10, 18

Rao, C.R. 45–47, 53
Richard, J.-F. 3–18
Robinson, A.G. 67, 71, 82, 97, 101–103
Robinson, S. 105
Rockett, K. 138, 139, 154
Rockler, N. 104
Rogers, J. 19–32
Round, J.I. 70, 106

Samuelson, P.A. 67, 106
Schnabel, R.B. 28
Schwartz, N.L. 159, 163
Schwartz, T. 29
Shea, B.L. 33, 42
Sims, C.A. 56 62
Singleton, K.J. 6
Sisler, M. 109, 120
Stephan, F.F. 67, 72, 105
Stone, R. 67, 71, 106
Stuart, A. 21
Svensson, L.E.O. 159, 163
Swamy, P.A.V.B. 43–54

Takayama, T. 67, 106
Taylor, J.B. 113, 119
Telly, H. 21
Theil, H. 57, 61, 62
Tinsley, P.A. 43–54
Tobler, W. 104
Trivedi, P.K. 14, 18
Tse, E. 108, 111–115, 120
Tsitsiklis, J.N. 86, 105
Tukey, J.W. 69, 106

Van der Ploeg, F. 71, 106
Van der Sluis, A. 66, 106
Van Velden, L.M.T. 157–163
Vanderbilt, D. 21, 29
Venables, A.J. 157, 163
Verdier, F. 105

Wallis, K.F. 113, 120
Wasserman, P.F.D. 21, 29
Wong, D.F. 21, 29

Yancey, T.A. 69, 105
Young, D.M. 109, 120

Zikan, K. 86, 105

Advanced Studies in Theoretical and Applied Econometrics

1. J.H.P. Paelinck (ed.): *Qualitative and Quantitative Mathematical Economics.* 1982 ISBN 90-247-2623-9
2. J.P. Ancot (ed.): *Analysing the Structure of Econometric Models.* 1984 ISBN 90-247-2894-0
3. A.J. Hughes Hallet (ed.): *Applied Decision Analysis and Economic Behaviour.* 1984 ISBN 90-247-2968-8
4. J.K. Sengupta: *Information and Efficiency in Economic Decision.* 1985 ISBN 90-247-3072-4
5. P. Artus and O. Guvenen (eds.), in collaboration with F. Gagey: *International Macroeconomic Modelling for Policy Decisions.* 1986 ISBN 90-247-3201-8
6. M.J. Vilares: *Structural Change in Macroeconomic Models.* Theory and Estimation. 1986 ISBN 90-247-3277-8
7. C. Carraro and D. Sartore (eds.): *Development of Control Theory for Economic Analysis.* 1987 ISBN 90-247-3345-6
8. D.P. Broer: *Neoclassical Theory and Empirical Models of Aggregate Firm Behaviour.* 1987 ISBN 90-247-3412-6
9. A. Italianer: *Theory and Practice of International Trade Linkage Models.* 1986 ISBN 90-247-3407-X
10. D.A. Kendrick: *Feedback.* A New Framework for Macroeconomic Policy. 1988 ISBN 90-247-3593-9; Pb: 90-247-3650-1
11. J.K. Sengupta and G.K. Kadekodi (eds.): *Econometrics of Planning and Efficiency.* 1988 ISBN 90-247-3602-1
12. D.A. Griffith: *Advanced Spatial Statistics.* Special Topics in the Exploration of Quantitative Spatial Data Series. 1988 ISBN 90-247-3627-7
13. O. Guvenen (ed.): *International Commodity Market Models and Policy Analysis.* 1988 ISBN 90-247-3768-0
14. G. Arbia: *Spatial Data Configuration in Statistical Analysis of Regional Economic and Related Problems.* 1989 ISBN 0-7923-0284-2
15. B. Raj (ed.): *Advances in Econometrics and Modelling.* 1989 ISBN 0-7923-0299-0
16. A. Aznar Grasa: *Econometric Model Selection.* A New Approach. 1989 ISBN 0-7923-0321-0
17. L.R. Klein and J. Marquez (eds.): *Economics in Theory and Practice.* An Eclectic Approach. Essays in Honor of F. G. Adams. 1989 ISBN 0-7923-0410-1
18. D.A. Kendrick: *Models for Analyzing Comparative Advantage.* 1990 ISBN 0-7923-0528-0
19. P. Artus and Y. Barroux (eds.): *Monetary Policy.* A Theoretical and Econometric Approach. 1990 ISBN 0-7923-0626-0
20. G. Duru and J.H.P. Paelinck (eds.): *Econometrics of Health Care.* 1990 ISBN 0-7923-0766-6
21. L. Phlips (ed.): *Commodity, Futures and Financial Markets.* 1991 ISBN 0-7923-1043-8
22. H.M. Amman, D.A. Belsley and L.F. Pau (eds.): *Computational Economics and Econometrics.* 1991 ISBN 0-7923-1287-2

Kluwer Academic Publishers – Dordrecht / Boston / London

GPSR Compliance

The European Union's (EU) General Product Safety Regulation (GPSR) is a set of rules that requires consumer products to be safe and our obligations to ensure this.

If you have any concerns about our products, you can contact us on ProductSafety@springernature.com

In case Publisher is established outside the EU, the EU authorized representative is:

Springer Nature Customer Service Center GmbH
Europaplatz 3
69115 Heidelberg, Germany

Batch number: 10761599

Printed by Printforce, the Netherlands